Environmental Science and Engineering

The ultimate goal of this series is to contribute to the protection of our environment, which calls for both profound research and the ongoing development of solutions and measurements by experts in the field. Accordingly, the series promotes not only a deeper understanding of environmental processes and the evaluation of management strategies, but also design and technology aimed at improving environmental quality. Books focusing on the former are published in the subseries Environmental Science, those focusing on the latter in the subseries Environmental Engineering.

Qiang Wang · Ahmad Hosseini-Bandegharaei
Editors

Environmental Functional Materials in Air Pollution

Selected paper in 2024 9th International Conference on Advances in Energy and Environment Research

Springer

Editors
Qiang Wang
College of Environmental Science
and Engineering
Beijing Forestry University
Beijing, China

Ahmad Hosseini-Bandegharaei
Faculty of Chemistry
Semnan University
Semnan, Iran

ISSN 1863-5520 ISSN 1863-5539 (electronic)
Environmental Science and Engineering
ISBN 978-3-031-84570-3 ISBN 978-3-031-84571-0 (eBook)
https://doi.org/10.1007/978-3-031-84571-0

This Springer imprint is published by the registered company Springer Nature Switzerland AG
The registered company address is: Gewerbestrasse 11, 6330 Cham, Switzerland

Preface

On behalf of the conference organizing committee, we are deeply honored to write a preface for the proceedings of the 2024 9th International Conference on Advances in Energy and Environment Research (ICAEER 2024) to show you the wonderful moments and fruitful results of the conference.

Since the first International Conference on Advances in Energy and Environment Research was successfully held in 2016, it has become an important platform for domestic and foreign experts and scholars working in the field of energy and environment to exchange ideas and discuss cutting-edge science and technology. Over the past nine years, we have always adhered to the tenet of "innovation, cooperation and development" and are committed to promoting the progress and development of energy and environmental science. This conference, as the ninth event, brings together top scholars and industry elites from all over the world to discuss the latest research results and future development trends of energy and environment.

The conference is sponsored by Shanghai University of Electric Power and organized by College of Energy and Mechanical Engineering of Shanghai University of Electric Power. According to statistics, the conference attracts more than 200 experts and scholars from China, Canada, Australia, Malaysia, Bangladesh, Thailand and other countries, as well as teachers and students, which fully reflects the internationalization of the conference and extensive influence.

During the three-day conference, we arranged a rich and colorful agenda. The Honorary Chairmen of the conference were Prof. Junfu Lv from Tsinghua University (Academician of the Chinese Academy of Engineering) and Prof. Hualin Wang from East China University of Science and Technology (Academician of the Chinese Academy of Engineering), while the Chairs of the conference were Prof. Mingyan Gu from Anhui University of Technology and Prof. Jiang Wu from Shanghai University of Electric Power, China (top 2% scientists around the world). In addition, we have invited many famous scholars from home and abroad as keynote speakers, including Prof. Huiyan Zhang from Southeast University, China, Prof. Marc A. Rosen, member of the Canadian Academy of Engineering, Prof. Ahmad Zuhairi Abdullah, member of the Academy of Sciences of Malaysia, Prof. Mohammed J. K.

Bashir from Central Queensland University, Australia, etc. Their dramatic speeches added a strong academic atmosphere to the conference.

The keynote speeches of the conference cover a wide range of fields such as energy engineering and technology, environmental science and engineering, solid waste management, resource exploration and utilization, and sustainable development, aiming to provide a platform for experts and scholars in energy and environment engineering around the world to exchange their ideas and share their experiences. In addition, the conference also organized interdisciplinary forums in the fields of environmental science and environmental engineering, electrical energy engineering and energy technology, providing participants with a broader space for academic exchange. Participants expressed that, through this conference, they not only understood the latest development of the industry, but also met many like-minded experts and scholars, laying a solid foundation for future cooperation and exchange.

All the papers of this conference have been rigorously evaluated and finally included in the proceedings of the conference. These contributions not only show the latest research results in the field of energy and environment, but also provide valuable references for researchers in related fields. We believe that the publication of these papers will further promote the development and progress of energy and environmental science.

Finally, we would like to express our sincere thanks to all the experts and scholars for their great support and active participation in this conference! Let us work together to create a bright future for energy and environmental sciences!

The Committee of ICAEER 2024.

Beijing, China — Qiang Wang
Semnan, Iran — Ahmad Hosseini-Bandegharaei

Committee Members

Honorary Chairmen

Prof. Junfu Lv, Tsinghua University, China
Prof. Hualin Wang, East China University of Science and Technology, China

Conference Chairs

Prof. Mingyan Gu, Anhui University of Technology, China
Prof. Jiang Wu, Shanghai University of Electric Power, China

Technical Program Committee Chairs

Prof. Naiqiang Yan, Shanghai Jiao Tong University, China
Prof. Jing Liu, Huazhong University of Science and Technology, China
Prof. Fuping Qian, Anhui University of Technology, China
Prof. Huaqiang Chu, Anhui University of Technology, China
Prof. Ping He, Shanghai University of Electric Power, China

Publication Chairs

Prof. Ahmad Hosseini-Bandegharaei, Semnan University, Iran
Prof. Qiang Wang, Beijing Forestry University, China
Prof. Minghou Xu, Huazhong University of Science and Technology, China
Prof. Jiang Wu, Shanghai University of Electric Power, China

Publicity Chair

Assoc. Prof. Shuo Yang, Shenyang Institute of Engineering, China

Local Organizing Committee

Prof. Xinyuan Zhu, Shanghai Jiaotong University, China
Prof. Yaji Huang, Southeast University, China
Prof. Ping Lu, Nanjing Normal University, China
Prof. Lin Dong, Nanjing University, China
Prof. Weidong Fan, Shanghai Jiaotong University, China
Prof. Dong Liu, Nanjing University of Science and Technology, China
Prof. Chunbo Wang, North China Electric Power University, China
Prof. Qiang Lu, North China Electric Power University, China
Prof. Boxiong Shen, Hebei University of Technology, China
Prof. Zhongzhu Qiu, Shanghai University of Electric Power, China
Prof. Hongbo Ren, Shanghai University of Electric Power, China
Prof. Tao Song, Nanjing Normal University, China
Prof. Binlin Dou, University of Shanghai for Science and Technology, China
Prof. Yufeng Duan, Southeast University, China

Technical Program Committee

Prof. Yun Qiao, Shanghai University, China
Assoc. Prof. Youjia Liang, Wuhan University of Technology, China
Assoc. Prof. Yifeng Xu, Wuhan University of Technology, China
Assoc. Prof. Cheng Peng, Shanghai University of Electric Power, China
Assoc. Prof. Tianran Ye, Anhui University of Technology, China
Assoc. Prof. Bingbing Qiu, Anhui University of Technology, China
Assoc. Prof. Junchao Xu, Anhui University of Technology, China
Dr. Yifeng Chen, Shanghai University of Electric Power, China
Dr. Mengyuan Hu, Shanghai University of Electric Power, China
Assoc. Prof. Rongkui Su, Central South University of Forestry and Technology, China
Prof. Yankun Jiang, Huazhong University of Science and Technology, China
Prof. Guangqian Luo, Huazhong University of Science and Technology, China
Assoc. Prof. Zijian Zhou, Huazhong University of Science and Technology, China
Prof. Xiaoyan Liu, Shanghai University, China
Assoc. Prof. Shijie Song, Xi'an University of Science and Technology, China
Prof. Zhibao Huo, Shanghai Ocean University, China

Prof. Xinsheng Zhang, East China University of Science and Technology, China
Prof. Zhihua Zhang, China University of Petroleum (East China), China
Assoc. Prof. Xinlu Han, China University of Petroleum (East China), China
Prof. Yu Wang, Chongqing University, China
Prof. Ke Wang, Hohai University, China
Prof. Sheng Chen, Hohai University, China
Assoc. Prof. Bo Zhang, North China Electric Power University, China
Assoc. Prof. Chen Zhang, Shanghai Jiao Tong University, China
Assoc. Prof. Linni Jian, South University of Science and Technology of China, China
Assoc. Research Fellow, Qiang Fu, Sichuan University, China
Assoc. Prof. Yongning Zhao, China Agricultural University, China
Assoc. Prof. Qiushi Cui, Chongqing University, China
Prof. Changjun Xie, Wuhan University of Technology, China
Assoc. Prof. Yue Yang, Hefei University of Technology, China
Assoc. Prof. Zhiqing Yang, Hefei University of Technology, China
Assoc. Prof. Yongxi Zhang, Changsha University of Science and Technology, China
Assoc. Prof. Wenlong Fu, China Three Gorges University, China
Prof. Zhizhao Che, Tianjin University, China
Prof. Xingyu Liang, Tianjin University, China
Prof. Cheng Huang, Soochow University, China
Assoc. Prof. Yanzhou Qin, Tianjin University, China
Assoc. Prof. Chao Li, Tianjin University, China
Assoc. Prof. Qian Xiao, Tianjin University, China
Researcher Long Jiang, Zhejiang University, China
Assoc. Research Fellow, Xiaojie Lin, Zhejiang University, China
Researcher, Peng Liu, Zhejiang University, China
Assoc. Prof. Xiang Li, Nantong University, China
Prof. Song Xiao, Linyi University, China
Prof. Buxiang Zhou, Sichuan University, China
Assoc. Researcher, Yiwei Qiu, Sichuan University, China
Assoc. Prof. Tianlei Zang, Sichuan University, China
Assoc. Research Fellow, Shuaijia He, Sichuan University, China
Prof. Jian Guo Zhu, The University of Sydney, Australia
Prof. Gordon Huang, University of Regina, Canada
Prof. Jianxun Jin, Tianjin University, China
Dr. Zuhaila Mat Yasin, Universiti Teknologi MARA, Malaysia
Assoc. Prof. Shamsul Aizam Zulkifli, Universiti Tun Hussein Onn Malaysia, Malaysia
Dr. Hamed Aly, Dalhousie University, Canada
Prof. Said Drid, University of Batna 2, Algeria
Assoc. Prof. Devendra Kumar Doda, Vivekananda Global University, India
Prof. Nava Raj Karki, Tribhuvan University, Nepal
Prof. Tianguang Lv, Shandong University, China
Prof. Yuanyuan Wang, Changsha University of Science and Technology, China
Assoc. Xinye Wang, Nanjing Normal University, China

Assoc. Prof. Hongyun Hu, Huazhong University of Science and Technology, China
Assoc. Prof. Sheng Chen, Huazhong University of Science of Technology, China
Assoc. Prof. Yao Zhao, Shanghai Jiao Tong University, China
Assoc. Prof. Yanqing Niu, Xi'an Jiaotong University, China
Prof. Gang Li, Dalian Jiaotong University, China
Prof. Qian Xu, Jiangsu University, China
Prof. Guohua Liu, North China Electric Power University, China
Prof. Alexandru Salceanu, Technical University of Iasi, Romania
Prof. Bodius Salam, Chittagong University of Engineering and Technology, Bangladesh
Dr. Yuanfan Zheng, Shanghai Normal University, China
Prof. Zheng Bo, Zhejiang University, China
Prof. Jinghong Zhou, Changchun Institute of Technology, China
Prof. Shyam S. Nandwani, Universidad Nacional Costa Rica, Costa Rica
Prof. Chun Yang, Nanyang Technological University, Singapore
Prof. Artem Khalatov, National academy of Sciences of Ukraine, Ukraine
Assoc. Prof. Dawan Wiwattanadate, Chulalongkorn University, Thailand
Dr. Abdeen Omer, Energy Research Institute (ERI), UK
Prof. Weiwei Yang, Xi'an Jiaotong University, China

Contents

Chapter 1
Study on Carbon-Ash Separation of Coal Gasification Fine Slag by Flotation with Composite Collector

Yuchen Xiao, Shiyong Zhao, and Zhenni Yang

Abstract Coal gasification fine slag (CGFS) produced by a Texaco coal water slurry gasifier in Shaanxi, China was taken as the research object. The carbon-ash separation of CGFS was carried out by flotation with composite collector. The raw material and the flotation product were characterized by XPS, BET, SEM–EDS, et al. analysis. The results showed that when the dosage of composite collector was 15 kg/t and MIBC was 4 kg/t, the flotation perfect index and the combustible recovery were 47.78 and 84.34%, respectively, which increased by 14.69 and 23.03% compared with dodecane as collector, but the concentrate ash content was not significantly reduced. XPS analysis shows that there are still a large number of hydrophilic groups on the surface of the unburned carbon in the tailings, and these hydrophilic groups hinder the mineralization process of the unburned carbon and collector, thus affecting the floating of residual carbon. BET and SEM analysis results showed that the specific surface area of the flotation concentrate was 303.50 m^2/g and the pore structure was developed, and EDS analysis showed that a large number of fine-grained spherical inorganic matter was still filled in the pore structure of unburned carbon, which was the main reason why the concentrate ash content was still about 30%.

Keywords Coal gasification fine slag · Flotation · Carbon-ash separation · Composite collector

1.1 Introduction

In China, a large amount of CGFS is produced in the process of coal gasification every year. Because there is no mature utilization method, most of CGFS is deposited or landfill, which not only wastes available resources, but also occupies land resources and pollutes the environment (Hu 2019; Gao et al. 2024). Therefore, the problem of

Y. Xiao (✉) · S. Zhao · Z. Yang
School of Chemistry and Chemical Engineering, Xi'an University of Science and Technology, Xi'an, China
e-mail: 2602898644@qq.com

Q. Wang and A. Hosseini-Bandegharaei (eds.), *Environmental Functional Materials in Air Pollution*, Environmental Science and Engineering,
https://doi.org/10.1007/978-3-031-84571-0_1

resource utilization of CGFS should be solved urgently. The studies show that there are rich silica, alumina, iron oxide and residual carbon in CGFS, but there is a mutual restriction between the utilization of inorganic material and residual carbon. Due to its high carbon content and high calorific value, residual carbon can be incorporated into boiler fuel for combustion, as well as its developed pore structure and high specific surface area, it can be used for the preparation of high value-added porous carbon materials, but inorganic minerals will reduce the heat energy conversion efficiency and porosity of fine slag (Zhang et al. 2023; Bin et al. 2024). At the same time, the ash in the fine slag can be mixed into some building materials to improve their compressive strength and economic benefits, but the presence of carbon residue makes its LOI less than the standard of building materials, thus limiting its utilization (Fu et al. 2020; Gu et al. 2021). Therefore, the pre-separation of carbon and ash in CGFS is the premise of realizing efficient resource utilization.

Froth flotation is one of the main methods for effective carbon-ash separation, which depends on the difference in surface hydrophobicity between unburned carbon and inorganic minerals in CGFS (Mondal et al. 2021; Guo et al. 2020). However, the traditional oil collectors (such as dodecane and kerosene) are not ideal for the flotation of CGFS, and the cost is high (Zhao et al. 2016; Ren 2021). Hu (2020) found that the surface of unburned carbon in gasification fine slag is rich in pores and easy to absorb water, which is not conducive to the adhesion between bubbles and particles. By adding carbon black to fill the pores of unburned carbon, the floatability of unburned carbon is improved. This suggests the existence of pores can affect flotation. Yu et al. (2022) measured the hydrogen proton signal of collectors in the pores by ^{1}H LF-NMR, and proved that both polar and non-polar agents could be absorbed into the pores, which resulted in the consumption of a large number of agents in the flotation of CGFS. Ma (2020) compounding non-polar hydrocarbon oil (dodecane) with lower ester oil (ethyl butyrate, ethyl caprylate, ethyl laurate). It was found that the combustible recovery of dodecane-ester composite collector was significantly higher than that of single collector. Dodecane is mainly adsorbed on the hydrophobic point of the coal surface to ensure selectivity, and ester molecules are mainly adsorbed on the hydrophilic point of the coal surface to improve the hydrophobicity of the coal surface, so as to promote the spread and adhesion of the collector on the coal surface to optimize the flotation effect.

In this paper, non-polar hydrocarbon oil dodecane and polar composite collector were used as collector to carry out carbon-ash separation flotation experiment of CGFS, and the physical and chemical properties of concentrate and tailings obtained by flotation were characterized by XPS, BET, SEM–EDS et al. analysis. It lays a theoretical foundation for the subsequent effective carbon-ash separation of the CGFS.

1.2 Materials and Methods

1.2.1 Materials

The CGFS was obtained from the coal water slurry gasifier of a coal chemical plant in Shaanxi, China. The results of raw material proximate and elemental analysis are shown in Table 1.1. The chemical composition analysis is shown in Table 1.2. The particle size composition analysis is shown in Fig. 1.1, and the mineral composition of the sample is shown in Fig. 1.2.

It can be seen from Table 1.1 that the ash content of CGFS is 53.60%, the fixed carbon content is 42.63%, and the sulfur content is low (0.42%), which indicates that the unburned carbon with a high content in the CGFS can be recycled through sorting. In Fig. 1.1, the particle size < 0.045 mm was the dominant particle size, with a yield of 37.76% and ash content of 75.69%, indicating that the sample contained a large amount of fine ash. Table 1.2 is found that the inorganic minerals in CGFS are mainly silicon oxides and hydrophilic minerals, which is consistent with the results

Table 1.1 The proximate and elemental analysis of CGFS

Proximate analysis (wt.%)				Elemental analysis (wt.%)				
M_{ad}	A_d	V_{daf}	FC_d	N_{ad}	C_{ad}	H_{ad}	O_{ad}	$S_{t,ad}$
1.18	53.60	7.02	42.63	0.832	43.96	0.488	0.18	0.42

Table 1.2 The chemical composition of CGFS (wt.%)

SiO_2	Al_2O_3	CaO	Fe_2O_3	TiO_2	Na_2O	Others	LOI
31.92	14.42	3.60	2.95	0.77	0.71	1.24	44.39

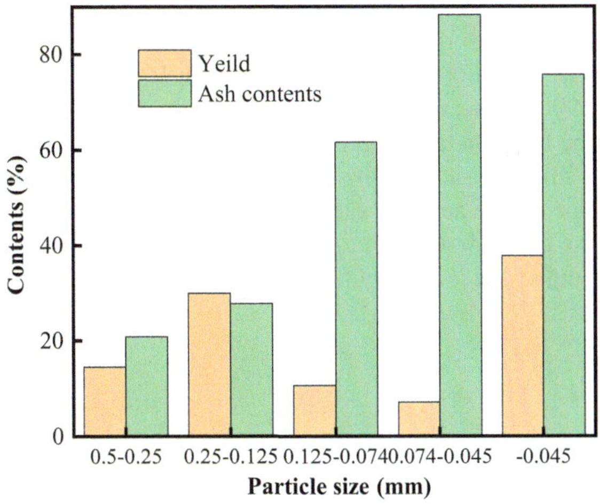

Fig. 1.1 The particle size composition of CGFS

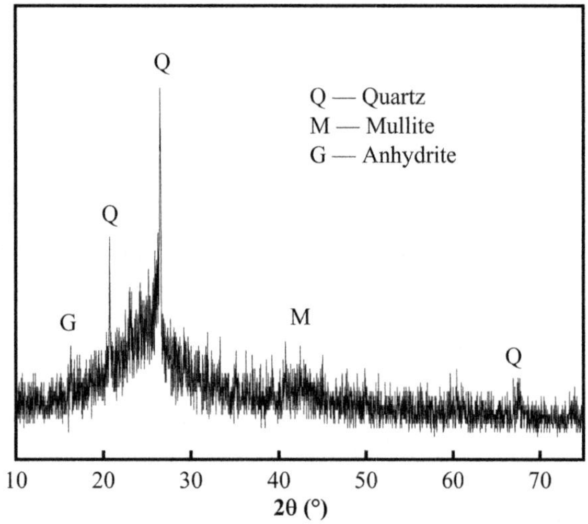

Fig. 1.2 The mineral composition of CGFS

of XRD analysis (Fig. 1.2). Therefore, the hydrophobicity difference between the surface of unburned carbon and inorganic minerals can be used to realize carbon-ash separation by flotation.

In this experiment, the composite collector is composed of 60% dodecane ($CH_3(CH_2)_{10}CH_3$, Weng Jiang) and 40% dodecanoic acid ($CH_3(CH_2)_{10}COOH$, Macklin). MIBC ($(CH_3)_2CHCH_2CH\ (OH)CH_3$, Aladdin) is selected as the foaming agent.

1.2.2 Methods

1.2.2.1 Froth Flotation

The 1.0 L flotation cell was used, the pulp density was 60 g/L, the spindle speed was 1900 r/min, and the aeration rate was 0.15 m^3/h. The weighed CGFS was added into the flotation tank to prepare the pulp, stirred for 2 min after full wetting, stirred for 1 min after adding the collector, stirred for 1 min after adding the foaming agent, opened the aeration valve and the scraper to scrape the foam for 3 min, and finally the concentrate and tailings were collected. The flotation perfect index (η) and the combustible recovery (ε) were calculated using Eqs. (1.1) and (1.2), respectively:

$$\eta = \frac{\gamma_j}{100 - A_y} \times \frac{A_y - A_j}{A_y} \times 100\% \quad (1.1)$$

$$\varepsilon = \frac{\gamma_j(100 - A_j)}{(100 - A_y)} \times 100\% \quad (1.2)$$

where γ_j is the yield of concentrates (%), A_y is the ash content of feed (%), A_j is the ash content of concentrates (%).

1.2.2.2 Characterization of Samples

Surface elements of the feed, concentrate and tailings of flotation were examined by X-ray photoelectron spectroscopy (XPS). The BET specific surface areas and pore structure of the samples were analyzed using a Micrometrics ASAP 2460 N_2 adsorption analyzer. The surface morphology and element content in microscopic regions of samples was characterized by scanning electron microscope and energy disperse spectroscopy (SEM–EDS).

1.3 Results and Discussion

1.3.1 Flotation Experiment Results

Figure 1.3 shows the experimental results of flotation of CGFS with different of collector, the MIBC dosage was maintained at 4 kg/t.

No matter dodecane or composite collector is used as collector, the combustible recovery and flotation perfect index of concentrate continue to increase with the increase of the amount of collector, and the ash content of concentrate continues to decrease. However, compared with dodecane, the flotation with the composite collector of dodecane and dodecanoic acid can obtain better separation effect of carbon and ash. When the dosage of composite collector increased from 7 to 15 kg/t, the combustible recovery increased from 50.09 to 84.34%, the ash content of concentrate decreased from 37.17 to 32.63%, and the flotation perfection index increased from 24.93 to 47.78%. When the dosage of reagent is the same 15 kg/t, the combustible recovery of composite collector is 23.03% higher than that of dodecane, the flotation perfect index is 14.69% higher, but the ash content of concentrate is only 0.82% lower, indicating the use of composite agents as collector, the concentrate yield is greatly improved, but the ash content of concentrate is not changed much. Therefore, it is necessary to further analyze the physical and chemical characteristics of the concentrate and tailings products after the separation of the composite collector, explore the reasons why the ash content of the flotation concentrate is not significantly reduced, and provide a theoretical basis for the efficient separation of carbon ash.

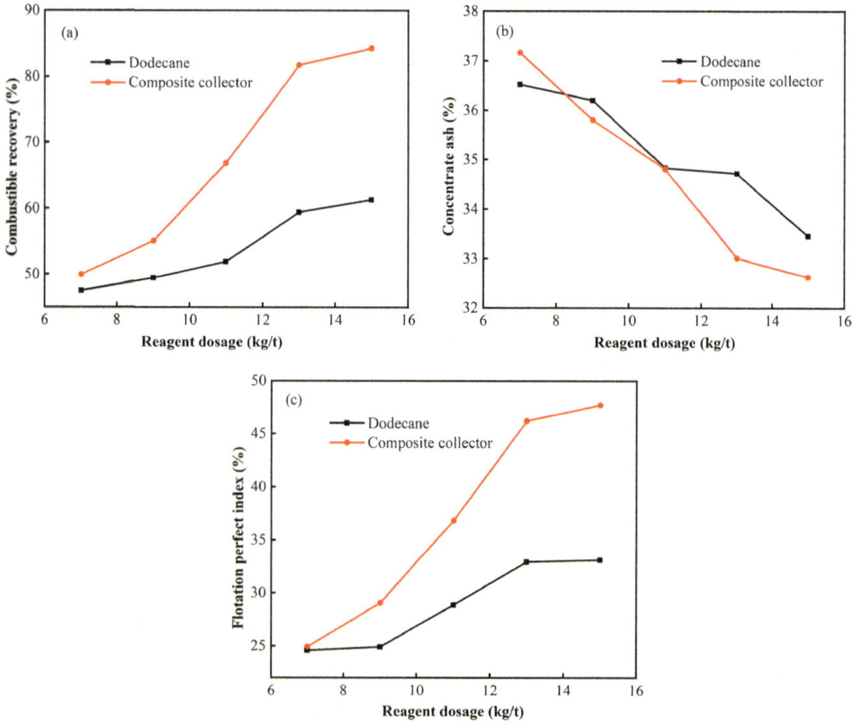

Fig. 1.3 Effect of flotation of CGFS with different of collector

1.3.2 Proximate Analysis

The proximate analysis results of CGFS and flotation products are shown in Table 1.3. The fixed carbon content in the raw material is 42.63%, and the dry base ash content of the concentrate after separation is 32.63%, indicating that there are still some inorganic substances in the concentrate that cannot be separated, and there are 14.27% carbon content in the tailings products, which still cannot meet the requirements of direct building materials. Therefore, the particle size analysis of concentrate and tailings is further carried out to explore the yield and ash composition of each particle grade product.

Table 1.3 The proximate analysis results of CGFS and flotation products

	M_{ad} (%)	A_d (%)	V_{daf} (%)	FC_{ad} (%)
Raw material	1.18	53.60	7.02	42.63
Concentrate	1.85	32.63	8.33	60.12
Tailings	0.55	81.91	20.71	14.27

Table 1.4 The particle size composition of CGFS and flotation products

Particle size (mm)	Raw material		Concentrate		Tailings	
	Yield (%)	Ash%	Yield (%)	Ash%	Yield (%)	Ash%
0.5–0.25	14.52	20.82	15.35	13.30	15.81	33.37
0.25–0.125	29.99	27.81	40.04	21.17	16.95	52.70
0.125–0.074	10.61	61.59	9.36	31.79	13.41	94.11
0.074–0.045	7.13	88.27	2.42	43.43	14.34	99.70
< 0.045	37.76	75.69	32.83	52.97	39.49	99.53
Total	100.00	52.77	100.00	31.93	100.00	80.43

1.3.3 The Particle Size Composition Analysis

The particle size composition analysis of CGFS and flotation products is shown in Table 1.4. It can be seen that 0.25–0.125 mm particle size is the dominant particle size in the concentrate, with cumulative yield of 40.04% and cumulative ash content of 21.17%. Compared with tailings of the same particle size, the ash content of tailings is only 52.70%, which is close to the raw material ash, indicating that a small part of coarse grained carbon still remains in the tailings and it fails to emerge with the foam. As a result, the separation of carbon ash in tailings is not complete. The cumulative yield of tailings with particle size less than 0.125 mm is 71.24%, and the cumulative ash content is 98.71%, while the cumulative yield of concentrates with the same particle size is 44.61%, and the cumulative ash content is 48.01%. It is said that the carbon-ash separation in the fine particle is not thorough enough because a large part of high ash inorganic substances are included in the concentrate foam with the carbon.

1.3.4 XPS Analysis

The results of the percentage content of elements on the surface of the raw material, concentrate and tailings of CGFS are shown in Fig. 1.4 and Table 1.5. Compared with thc percentage content of elements of the three samples, the percentage content of C1s in concentrate is the highest, which is 79.14%, while the percentage content of O 1s, Si 2p and Al 2p in tailings is the highest. The results show that most of the inorganic gray matter containing O, Si and Al on the surface of CGFS can be effectively separated from unburned carbon by separation.

The changes of carbon structure groups on the surface of carbon residues in CGFS were studied by peaking C1s peaks of the three samples, as shown in Fig. 1.5. The peaks of binding energy are 284.8, 285.9, 286.8 and 289.6 eV respectively, corresponding to C–C/C–H, C–O–C/C–OH, C=O and O–C=O groups (Peng and Yanli 2024), and the relative contents of each functional group are calculated according

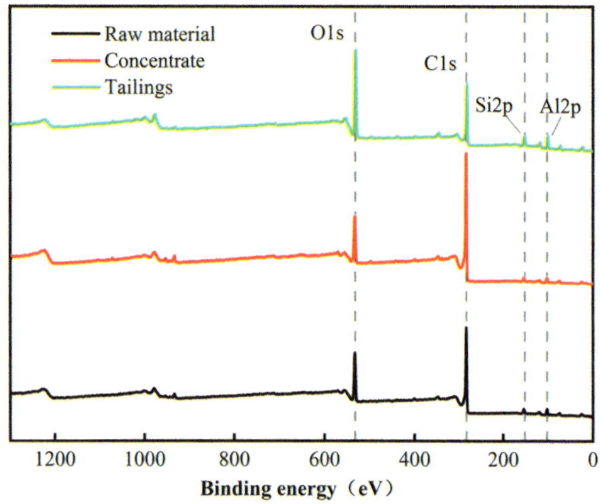

Fig. 1.4 XPS spectra of the samples

Table 1.5 The elements content of the samples

Element content (%)	C 1s	O 1s	Si 2p	Al 2p
Raw material	71.30	19.20	4.73	4.76
Concentrate	79.14	13.60	3.16	4.10
Tailings	50.01	33.62	10.14	6.23

to the area of each peak, as shown in Table 1.6. The relative content of C–C/C–H is 56.03%, C–O is 13.94%, C=O is 11.01%, and O–C=O is 19.02%. Among them, the C–C/C–H group is a hydrophobic group, and the content of this group can represent the hydrophobicity of the sample to a certain extent, and the oxygen-containing group increases its surface hydrophilicity by increasing the hydrophilic site on its surface. Therefore, a large number of hydrophilic oxygen-containing functional groups on the surface of carbon residue in CGFS lead to the difficulty of flotation and poor floatability of CGFS. It can be seen that hydrophobic groups C–C/C–H on the surface of concentrate products increased after flotation, while oxygen-containing groups C–O–C/C–OH and C=O on the surface of tailings increased, indicating that the collector masked some oxygen-containing functional groups on the surface of carbon residues, thereby increasing its hydrophobicity and improving flotation efficiency.

1.3.5 BET Analysis

Due to the abundant pore structure produced by CGFS in the process of coal gasification, it is also the reason that leads to the poor hydrophobicity of the surface of the coal gasification fine slag, so it is necessary to add a large amount of hydrocarbon oil collector to form an effective mineralization. The pore structure parameters of

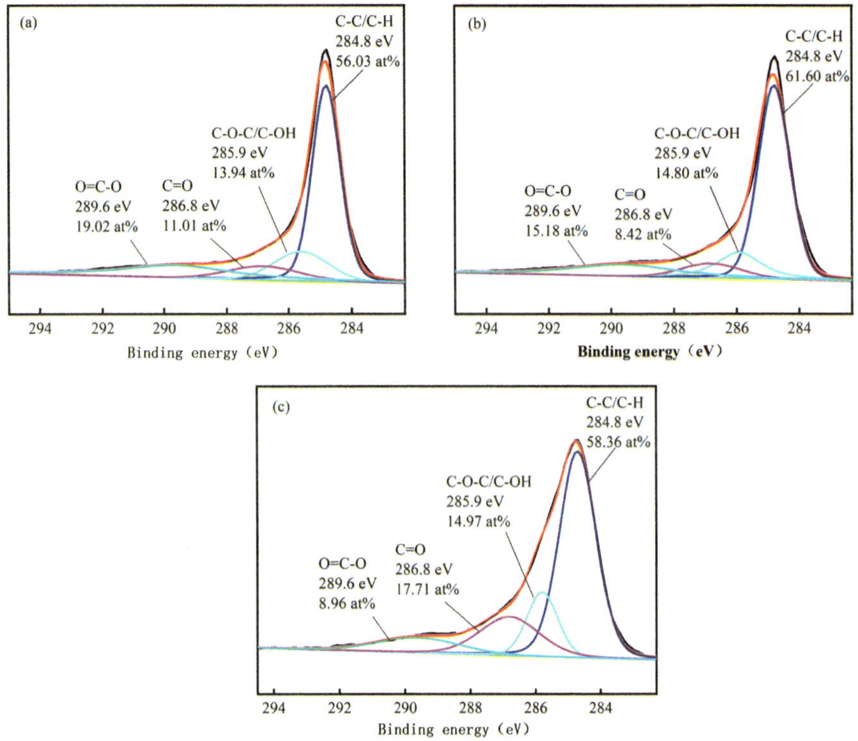

Fig. 1.5 XPS spectra in C 1s region of raw material (**a**), concentrate (**b**) and tailings (**c**)

Table 1.6 Each functional group contents of the samples C 1s region

Carbon structure content (%)	C–C/C–H	C–O–C/C–OH	C=O	O–C=O
Binding energy (eV)	284.8	285.9	286.8	289.6
Raw material	56.03	13.94	11.01	19.02
Concentrate	61.60	14.80	8.42	15.18
Tailings	58.36	14.97	17.71	8.96

the raw material and flotation products are shown in Table 1.7. The particle surface pore structure of the concentrate obtained through separation is the most developed, and its specific surface area and pore volume are the largest, with the specific surface area up to 303.50 m^2/g, and the total pore volume is 0.3311 cm^3/g, and the porosity is greatly different from that of the tailings products. Figure 1.6a shows the N_2 adsorption/desorption isotherms of the raw material and flotation products, all of which are typical IV curves, indicating the presence of a large number of mesoporous pores in the sample. When the relative pressure is in the range of 0–0.4, the multilayer adsorption occurs. When the relative pressure exceeds 0.4, the capillary condensation phenomenon of N_2 begins (Xue 2022). The adsorption capacity of concentrate

Table 1.7 Pore structure parameters of samples

Sample	BET specific surface area (m^2/g)	Total pore volume (cm^3/g)	Micropore volume (cm^3/g)	Average pore diameter (nm)
Raw material	203.27	0.2187	0.0164	4.30
Concentrate	303.50	0.3311	0.0150	4.36
Tailings	63.25	0.0678	0.0108	4.29

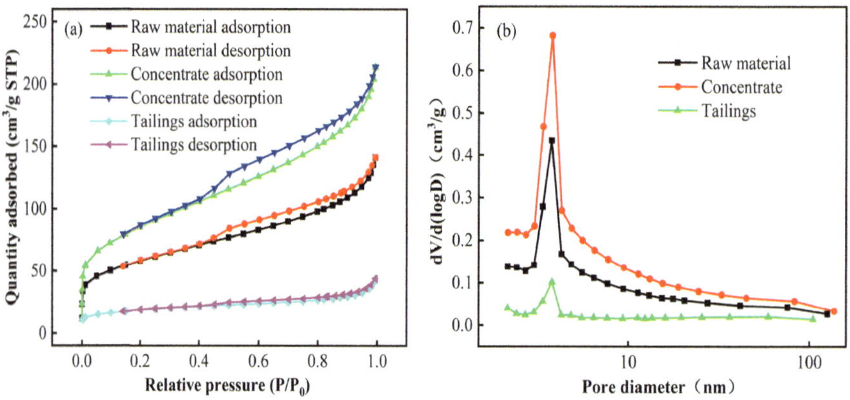

Fig. 1.6 Isothermal adsorption curve of N_2 (**a**) and pore size distribution (**b**)

products increased the most, indicating that the mesoporous distribution was the most uniform. Pore size distribution is shown in Fig. 1.6b, which is consistent with the results obtained by N_2 adsorption/desorption isotherm. Pore size of concentrate products is the most developed, and the average pore size of the raw material, concen trate and tailings products are about 4.30 nm, indicating that the pore structure of residual carbon distributed inconcentrate and tailings are basically the same. Therefore, it is believed that the well-developed pore structure of refined carbon makes it difficult for coarse carbon to emerge, and it is easy to absorb a large amount of water and hydrophobic agents into the pore structure, which makes the dosage of reagent consumed in the flotation process and leads to the fine ash being easily included in the foam to emerge.

1.3.6 SEM–EDS Analysis

Figure 1.7a is SEM image of the raw material, and it can be obviously observed that the irregular bulk, flake surface rough and porous unburned carbon and smooth spherical inorganic matter, they are either completely separated or mutually attached. By element determination in the area observed under 7500 × lens, it can be clearly observed that a large number of fine-grained spherical inorganic gray matter containing Si and Al elements adhere to the surface of the unburned carbon matrix or fill in the pore structure.

As shown in the SEM diagram of the concentrate in Fig. 1.7b, compared with raw material, the spherical inorganic gray matter existing alone is significantly reduced, indicating that the inorganic mineral particles attached to the carbon surface can be separated by flotation method by utilizing the hydrophobicity difference of the particle surface (Zhang et al. 2021). However, through EDS analysis of the area observed under 7500 × lens, the C content was 81.98%, and it was found that most of the spherical particles filled in the pores of unburned carbon still existed, and only part of the spherical inorganic gray matter was removed, so the ash content of the obtained concentrate was still about 30%. Figure 7c is the SEM diagram of tailings, and it can be found that the tailings particles are mainly smooth spherical particles. Through elemental analysis of the observation area under 7500 × lens, it is found that the tailings also contain a small part of uncompleted carbon particles like floc and some carbon elements embedded in inorganic matter, and the C content is 20.66%. This is also the main reason why it is difficult to increase the ash content of tailings and achieve the standard of direct utilization of building materials (Li 2021).

1.4 Conclusions

(1) Compared with dodecane, the use of dodecane and dodecanoic acid composite collector for flotation can obtain better carbon-ash separation effect, but the ash content of concentrate changes little. When the dosage of reagent is the same 15 kg/t, the combustible recovery is 23.03% higher than that of dodecane, the flotation perfect index is 14.69% higher, but the ash content of concentrate is only 0.82% lower.

(2) The ash content of flotation concentrate is 32.63%, and the analysis of particle size composition shows that some fine grained inorganic substances in the concentrate are included in the foam and emerge with the concentrate. And there is 14.27% carbon content in the tailings products, because the coarse particles of unburned carbon failed to fully mineralize and still remain in the tailings.

(3) XPS analysis shows that there are still a large number of hydrophilic groups on the surface of the unburned carbon in the tailings products, and these hydrophilic groups hinder the mineralization process of the unburned carbon and chemicals, thus affecting the emergence of residual carbon. BET and SEM analysis show that the coal gasification fine slag flotation feed and flotation concentrate have large specific surface area and complex pore structure, which is the main reason for the large consumption of flotation decarburization agent. Combined with EDS analysis, it is shown that a large number of fine-grained spherical inorganic gray matter containing Si and Al elements is still filled in the unburned carbon pore structure after separation, so that the ash content of the obtained concentrate is still about 30%.

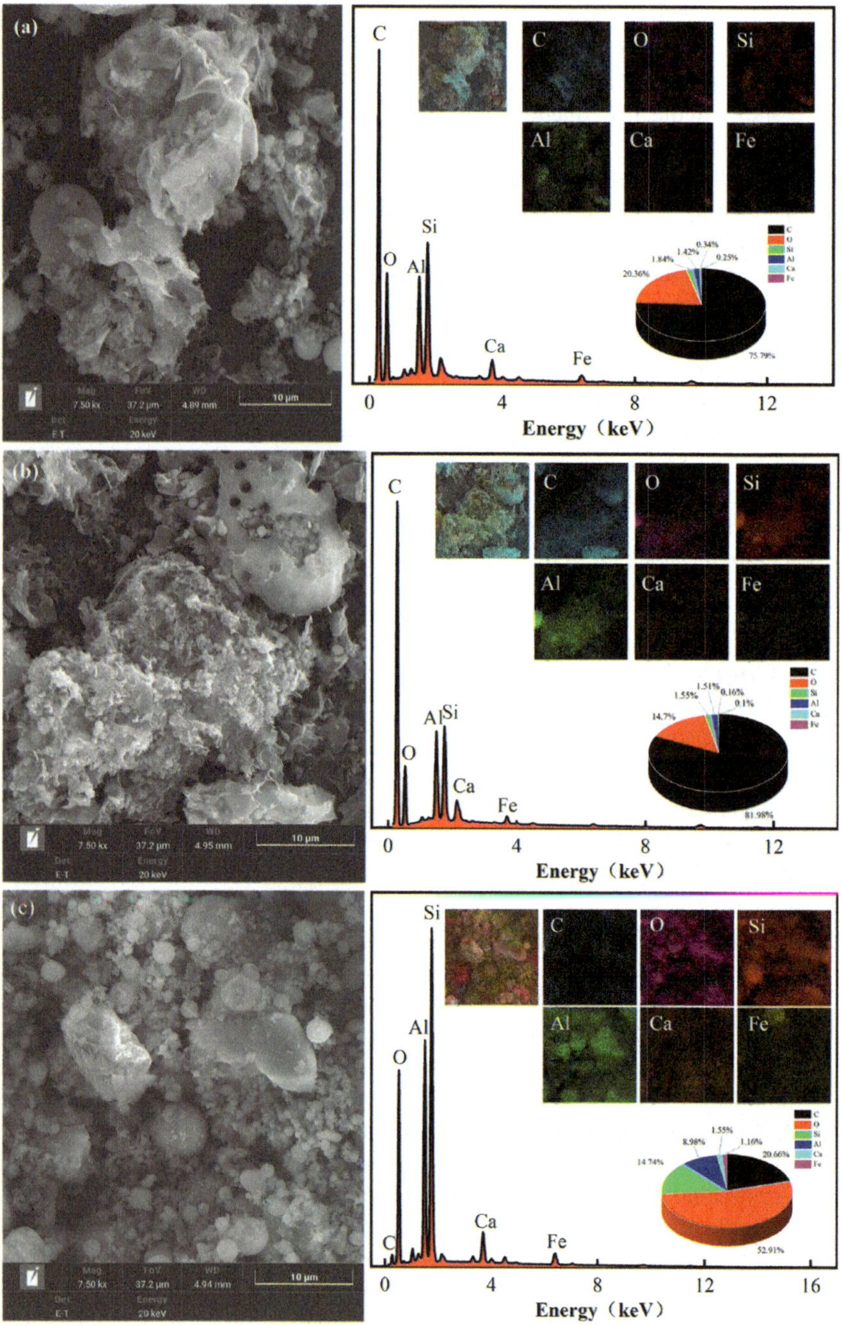

Fig. 1.7 SEM–EDS images of raw material (**a**), concentrate (**b**) and tailings (**c**)

References

Fu B, Ma M, Shen W et al (2020) Influence of coal gasification slag on strength and microstructure of Portland cement. Bull Chin Ceram Soc 39(08):2523–2527

Gao H, Liang L, Jin K et al (2024) Review on resource utilization of coal gasification slag. Coal Sci Technol 52(8):192–208

Gu Y, Wang D, Fang K et al (2021) Dissolution characteristics of coal gasification slag and its effect on cement-based materials. Bull Chin Ceram Soc 40(05):1579–1585

Guo F, Miao Z, Guo Z, Li J, Zhang Y, Wu J (2020) Properties of flotation residual carbon from gasification fine slag. Fuel 267:117043

Hu W (2019) Basic study on activation separation and resource utilization of Al-Si components in coal gasification residue. University of Chinese Academy of Sciences, Institute of Process Engineering, Chinese Academy of Sciences

Hu Z (2020) Study on the surface hydrophobicity improvement of unburned carbon in fly ash and gasification ash. China University of Mining and Technology

Li W (2021) Study on the mechanism and molecular simulation of compound reagents for gasification ash flotation. China University of Mining and Technology

Liu B, Wang Q, Chai Z et al (2014) Research on the formation mechanism and comprehensive utilization of entrained-flow coal gasification fine slag: progress and prospects. Clean Coal Technol 1–12. http://kns.cnki.net/kcms/detail/11.3676.TD.20231205.1007.005.html

Ma L (2020) Study on the attachment and spreading behavior and mechanism of compound collector at the low rank coal/water interface. China University of Mining and Technology

Mondal S, Acharjee A, Mandal U, Saha B (2021) Froth flotation process and its application. Viet J Chem 59(4):417–425

Peng W, Yanli L (2024) Effect of MO collector on coal gasification fine slag flotation. Multipurpse Util Mineral Resour 45(05):9–14

Ren P (2021) Experimental study on deashing of fine slag from coal gasification by grinding flotation. Xi'an University of Science and Technology

Xue Z (2022) Research on extraction of carbon from coal gasification fine slag using hydrophobic-hydrophilic separation. Taiyuan University of Technology

Yu W, Wang X, Liu L et al (2022) Study on flotation behavior of high carbon coal gasification fine slag. J China Coal Soc 47(S1):265–275

Zhang Y, Guo Y, Wang R et al (2021) Study on the physicochemical properties of Ningdong Coal gasification fine slag and its carbon-ash separation products. J China Coal Soc 46(S2):1096–1104

Zhang R, Liu D, He H et al (2023) Comprehensive utilization of coal gasification fine slag and carbon ash separation technology. Coal Eng 55(05):175–182

Zhao S, Wu Y, Li B (2016) Study on residue features and decarbonization of Texaco entrained flow gasifier. Coal Eng 48(09):29–32

Chapter 2
Molecular Dynamics Simulation of the Sorption Behavior of Organic Pollutants by Organo-Montmorillonite

Lan Zhang, Qiu Shen, and Tao Ding

Abstract Sorption is one of the main mechanisms by which montmorillonite is used in water pollution treatment. In this study, montmorillonite was used to adsorb phenolic pollutants in wastewater and to study the adsorption mechanism of the process. The sorption was analyzed for surfactants ($CTMA^+$) and different pollutants (phenol/2-naphthol). It was found that in the aqueous environment, water molecules had some influence on the arrangement of the components including modifiers, inorganic cations and contaminants, and the above components were slightly more active than in the dry state. The modifiers arrangement shifted from a laminar to a tilted distribution with the increase of loading amount, and the main sorption sites for organic pollutants were shifted from the inner surface of organo-montmorillonite to the organic phase in the organo-montmorillonite. The free binding energy of phenol under MMT-1.0 loading is − 211.780, which is slightly lower than − 207.066 and − 204.147 under MMT-1.75 and MMT-0.25 loading. When 2-naphthol is used, the free binding energy of MMT-1.0 reaches − 343.38, which is much lower than that of − 272.51 and − 217.92 under the loading of MMT-0.25 and MMT-1.75, indicating that the adsorption effect of organic matter under the loading of MMT-1.0 is better and the adsorption effect of montmorillonite on 2-naphthol is stronger than that of phenol.

Keywords Molecular dynamics simulation · Organo-montmorillonite · Sorption · Partition

---

L. Zhang
Ecological and Environmental Monitoring Center of Zhejiang Province, Hangzhou, China

Q. Shen · T. Ding (✉)
China Jiliang University, Hangzhou, China
e-mail: dingtao@cjlu.edu.cn

Q. Wang and A. Hosseini-Bandegharaei (eds.), *Environmental Functional Materials in Air Pollution*, Environmental Science and Engineering,
https://doi.org/10.1007/978-3-031-84571-0_2

2.1 Introduction

Organic pollutants and contamination may cause irreparable damage to water, and pose a threat to human health. Thus, it is necessary to effectively remove unwanted organic pollutants from the water system. Among the adsorbent material, montmorillonite is considered to be the most representative because of its excellent cation exchange capacity (CEC) and space modifiability (Shi et al. 2023; Spiridon et al. 2023; Khan et al. 2023). When montmorillonite is widely used as an environmental remediation material to adsorb organic pollutants, its adsorption performance is closely related to the microstructure of the interlayer domain of montmorillonite, which mainly refers to the space between the structural layers (Harouache et al. 2024).

From the available studies, the sorption effect of organo-montmorillonite is closely related to the material of the modifier and the arrangement of the modifier in the interlayer (Perera et al. 2023).

2.2 Model Construction and Calculation Method

We employed GROMACS software (Spoel et al. 2005) for the Molecular Dynamics (MD) simulation. Two montmorillonite slabs were generated to construct the interlayer space. The initial configuration of montmorillonite was built that every four Al^{3+} in the octahedral layer was replace by one Mg^{2+} and the isomorphic substitutions obey Loewenstein's rule. Na^+ was used to neutralize and balance the negative charge generated by the lattice substitution. Depending on the type of montmorillonite used, the unit-cell of studied Na-Mt was: $Na_{0.75}Al_{3.25}Mg_{0.75}Si_8O_{20}(OH)_4$. In this study, the cationic surfactants involved in the use was cetyltrimethylammonium ion ($CTMA^+$). In addition, to achieve the model charge balance, Na^+ ions as well as Br^+ ions were present in the organic montmorillonite model for modifier loadings less than or greater than 1.0 CEC, respectively. In molecular dynamics simulations, the interaction forces of each molecule are described by the molecular force field.

2.3 Results and Discussions

2.3.1 CTMA-Modified Montmorillonite

Density Distribution of the Modifier

The density distribution of each modifier within the interlayer domain of organo-montmorillonite and snapshots are shown in Figs. 2.1 and 2.2. After running MD simulation, $CTMA^+$ showed aligned parallel to the clay surface, and covered the oxygen plane when the modifier loading on the clay was low. In the case of low loading (0.25CEC), C on the alkyl chains showed peaks on both sides and in the middle of the density distribution diagram. Combined with the snapshot of the interlayer domain of organo-montmorillonite, it can be found that the alkyl chains mainly show a bilayer distribution, and it is presumed that the peak in the middle is due to some bending and overlapping of the alkyl chain length. When the loading amount of CTMA cations in the montmorillonite are half of its cation exchange capacity (CEC), the density profile of alkyl chain shows two obvious peaks, which means CTMA is distributed bilaterally in the interlayer. With the growing amount of cation loading (1.0CEC), the organic aggregates are formed by intertwined alkyl chains in this aqueous environment. The arrangement of the alkyl chains changes from a layered distribution to an inclined distribution.

Meanwhile, the density profile of the N atom shows two peaks close to the siloxane surface, which means that the hydrophilic end of the CTMA is attached to the siloxane surface. The density profiles of Na atoms are also observed in two peaks, and both peaks are quite close to the surface of montmorillonite. Similar to the peaks of the N atoms, some small peaks are observed in the density profile of Br ions which indicates that they are quite close to each other by electrostatic force. Besides, the

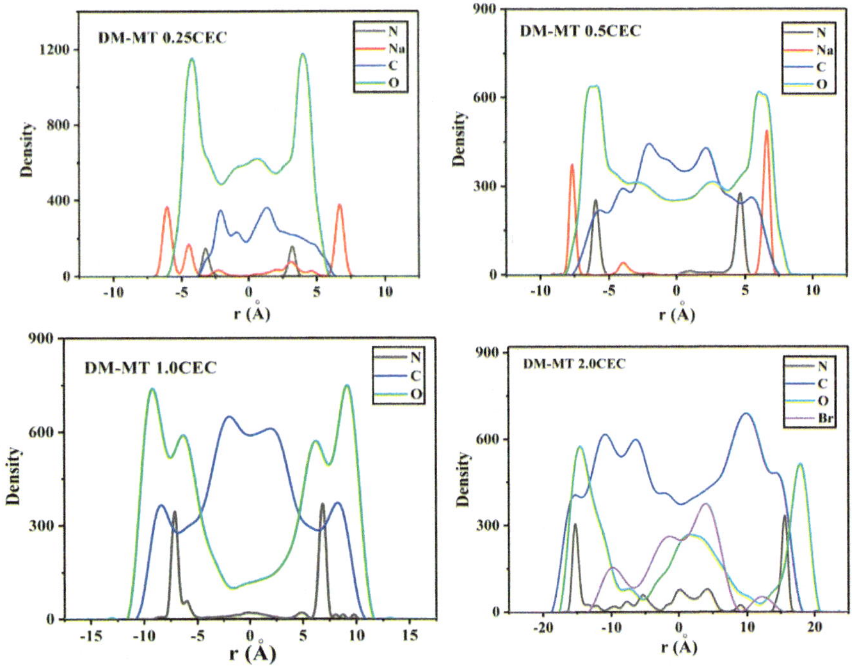

Fig. 2.1 The *z*-axis density distribution of CTMA-montmorillonite (black—ammonium N, red—Na + , blue—alkyl chain C, green—hydroxide O, purple—Br)

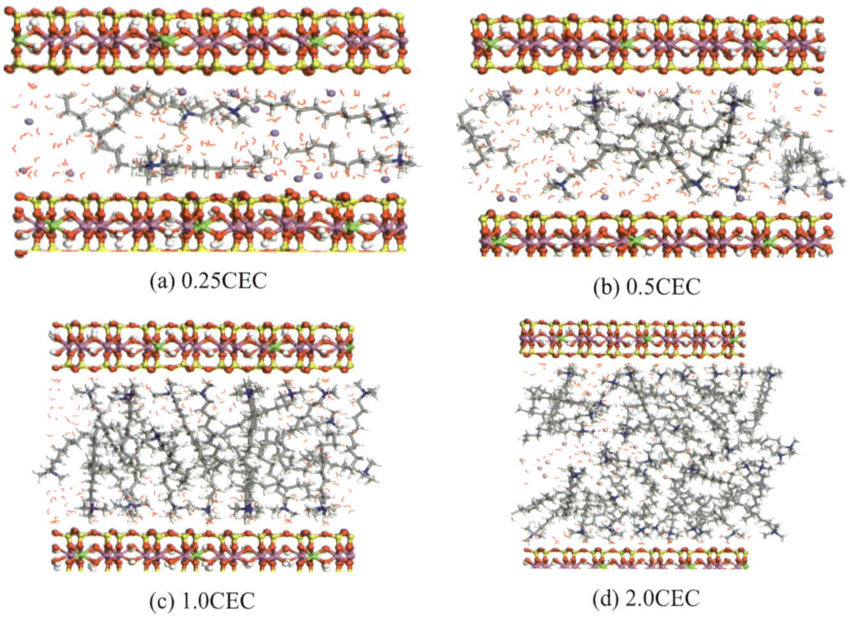

Fig. 2.2 Snapshot of CTMA montmorillonite (Purple—Na, bule—N, gray—C, red—O, brown—Br)

density profile of oxygen atoms on water molecules (O_w) shows that water molecules are mostly distributed on both sides near siloxane surface.

With the loading amount of modifier further increases (2.0CEC), the density profile of the N atoms forms small peaks in microstructure, which demonstrated that N atoms gradually leave the surface of the montmorillonite and enter the microstructure center.

The density distribution of Br ions showed a bilayer distribution and was more consistent with the distribution of the interlayer ammonium group N. In contrast, the density distribution of the alkyl chain C was misaligned with the Br ion.

2.3.2 *Sorption to the Organo-Montmorillonite*

Sorption Site of Organics to CTMA-Montmorillonite

The following Fig. 2.3 shows the snapshots and density distribution of CTMA-modified montmorillonite after sorption of small amounts of phenol at three different loading conditions. It can be seen that the entry of a small amount of phenol molecules into the interlayer domain does not affect the arrangement of CTMA alkyl chains.

When the loading amount of CTMA cations is 2.0 CEC, part of phenol molecules appeared around the siloxane surface again. In this circumstance, the interlayer

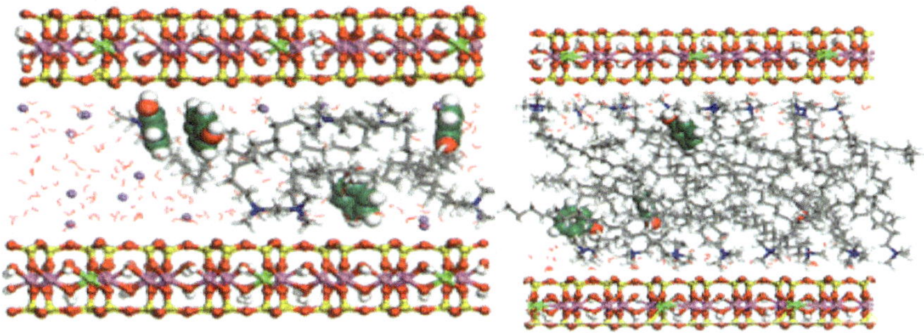

Fig. 2.3 Snapshot of CTMA montmorillonite with phenol (Purple—Na, bule—N, gray—C, red—O, brown—Br)

spacing is much larger than that of organoclay which is under low modifier loading. However, due to the existence of large number of organic active agent molecules, the mobility of the components in the layers is greatly reduced even with larger interlayer spacing, resulting in a steric hindrance effect which has a certain hindering effect on the sorption of phenol. The stacking density of CTMA is high, which inhibited the partition of phenol into the organic aggregate (Zhou et al. 2014).

The main sorption site of 2-naphthol molecules changed from the inner surface of montmorillonite to CTMA organic aggregates with the change of loading. When the CTMA loading is low, the sorption sites of 2-naphthol molecules are mainly on the inner silicone surface of montmorillonite. As can be seen from the snapshot, the -OH end of the 2-naphthol molecule is relatively close to the inner surface of the montmorillonite, and there is hydrogen bonding between the -OH end of the 2-naphthol molecule and the surface oxygen on the silicone surface, indicating the existence of coordination between the two. At a loading of 1.75 CEC, some 2-naphthol molecules are still present near the inner surface of montmorillonite, while the rest are sorbed in the organic aggregates of montmorillonite, this result is also consistent with the simulation snapshots from Fig. 2.4, which means that both surface adsorption and partitioning had a significant contribution to the sorption process. When the loading is much higher, the stacking density of CTMA between the layers increased, and there may be a spatial site resistance effect, which has a certain hindering effect on the sorption of 2-naphthol molecules. The experimental results showed that the best sorption effect was achieved when the loading level was 1.0 CEC.

Free Binding Energy

The binding energy between the ligand and the acceptor in the solvent can be defined by the following equation.

$$\Delta G_{binding} = G_{complex} - \left(G_{receptor} + G_{ligand}\right) \tag{2.1}$$

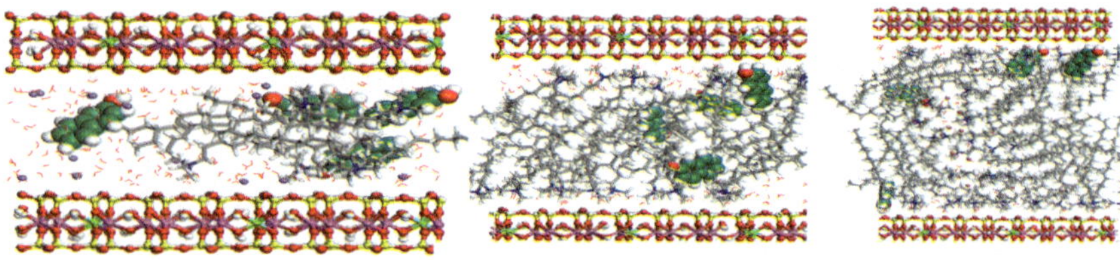

Fig. 2.4 Snapshot of CTMA montmorillonite with 2-naphthol (Purple—Na, bule—N, gray—C, red—O, brown—Br)

$$G_x = E_{MM} \pm TS + G_{solvation} \tag{2.2}$$

$$G_{solvation} = G_{polar} + G_{nonpolar} \tag{2.3}$$

$$G_{nonpolar} = \gamma A + b \tag{2.4}$$

In this study, phenol and 2-naphthol are considered as ligand and other components in the interlayer domain are jointly considered as an acceptor, and the calculation of binding free energy is carried out by the MM-PBSA method. Conformation was taken every 10 ps from the last 500 ps of the trajectory, a total of 50 conformations were used for the binding free energy calculation, and the average was taken as the result. The binding free energy of organo-montmorillonite with different modifier loadings for phenol sorption are listed in the Table 2.1.

From table, we can see all binding free energies are negative, indicating spontaneity of the reaction. For the sorption of phenol, the calculated results of binding free energy were as follows: MMT-1.0 > MMT-1.75 > MMT-0.25, it suggests that organo-montmorillonite with best sorption capacity towards phenol is MMT-1.0 followed by MMT-1.75 and MMT-0.25. The calculated results are consistent with the experimental results.

The results of the binding free energy of CTMA montmorillonite for the sorption of 2-naphthol and phenol, respectively, showed that the binding free energy values

Table 2.1 Binding free energy of phenol/2-naphthol interaction with modified montmorillonite

	Pollutant	$\Delta G_{ele}{}^{a}$	$\Delta G_{vdw}{}^{b}$	$\Delta G_{PB}{}^{c}$	$\Delta G_{NB}{}^{d}$	G_{bind}
MMT-0.25	Phenol	– 153.027	– 161.507	182.903	– 72.515	– 204.147
MMT-1.00		– 177.041	– 94.700	120.415	– 60.454	– 211.780
MMT-1.75		– 151.624	– 184.805	183.576	– 54.203	– 207.066
MMT-0.25	2-Naphthol	– 405.00	– 165.76	336.74	– 34.50	– 272.51
MMT-1.00		– 378.93	– 191.51	260.99	– 33.94	– 343.38
MMT-1.75		– 278.22	– 206.90	303.64	– 36.44	– 217.92

for the sorption of 2-naphthol were greater than those for phenol when the same loading of CTMA montmorillonite was used to adsorb the same amount of phenol and 2-naphthol. In other words, the sorption effect of CTMA montmorillonite for 2-naphthol was stronger than that for phenol.

2.4 Conclusion

The simulation results illustrate that the arrangement pattern of CTMA in the interlayer domain of montmorillonite changes from a bilayer lying distribution to a monolayer tilted distribution as the modifier loading changes. The sorption sites of phenol and 2-naphthol were concentrated on the surface at low loading. As the loading of $CTMA^+$ increases, the sorption sites shift to the organic phase formed by the entangled aggregation of $CTMA^+$. The negative value of the free energy change indicates that the sorption of contaminants on organo-montmorillonite is spontaneous. The analysis of the free energy magnitude revealed that the optimum sorption of the modifier occurred at 1.0 CEC and the sorption of 2-naphthol was better than that of phenol, indicating that the hydrophobic effect plays an important role in the sorption process. Organo-montmorillonite sorption of organic pollutants from the aqueous phase is mainly partitioning.

References

Harouache A, Kharroubi M, Lefkaier IK et al (2024) Adsorption of anionic dye onto homoionic montmorillonite: effect of the exchangeable cation on the adsorption. J Iran Chem Soc 21:1423–1438

Khan S, Ajmal S, Hussain T et al (2023) Clay-based materials for enhanced water treatment: adsorption mechanisms, challenges, and future directions. J Umm Al-Qura Univ Appl Sci

Perera MDR, Amarasena RALR, Bandara WMAT, Weerasooriya R, Jayarathna L (2023) Surfactant-modified clay composites: water treatment applications. In: Vithanage M, Lazzara G, Rajapaksha AU (eds) Clay composites. Advances in material research and technology. Springer, Singapore

Shi L, Liu W, Zhang X et al (2023) Adsorption of methylene blue from aqueous solution by crosslinked carboxymethyl cellulose/organo-montmorillonite composite hydrogels. J Polym Res 30:305

Spiridon I, Bele A, Apostol I et al (2023) Enhancing natural polymers-based materials using montmorillonite: preparation, characterization, and environmental applications. J Polym Environ

Spoel DVD, Lindahl E, Hess B et al (2005) GROMACS: fast, flexible, and free. J Comput Chem 26(16):1701–1718

Zhou Q, Shen W, Zhu J et al (2014) Structure and dynamic properties of water saturated CTMA-montmorillonite: molecular dynamics simulations. Appl Clay Sci 97–98:62–71

Chapter 3
Potential of Rice Husk Biochar in Cement Mortar for Carbon Capture and Storage

Priscilla Fong Ern Gunn, Chiu Chuen Onn, Hwei Voon Lee, and Kim Hung Mo

Abstract Research on green alternatives from agricultural waste as cement substitutes has been significant. Biochar, with unique properties, can improve cementitious composites' physical, mechanical properties and aiding carbon capture and storage. In this study, cement mortars were fabricated using rice husk biochar (RHB) at varying dosages - 10%, 20%, 30%, and 40% relative to the cement volume. The carbonation curing has been shown to greatly enhance the cement mortar's compressive strength, with a maximum increase of 82%. However, a higher dosage of biochar resulted in reduced workability and increased permeable pore space. This study indicates that biochar derived from rice husk waste has the potential to be utilized as a substitute in cement mortar, contributing to waste reduction and offers a promising solution for carbon capture and storage.

Keywords Carbon capture and storage · Waste valorization · Cementitious materials

3.1 Introduction

With urbanization and economic growth, the demand for infrastructure is increasing, leading to higher pollution with an approximate of 4.27 billion tons of cement is consumed annually. The cement industry, responsible for about 7% of global carbon

---

P. F. E. Gunn · C. C. Onn (✉)
Department of Civil Engineering, Faculty of Engineering, Universiti Malaya, Kuala Lumpur, Malaysia
e-mail: onnchiuchuen@um.edu.my

H. V. Lee
Nanotechnology and Catalysis Research Centre, Institute of Postgraduate Studies, Universiti Malaya, Kuala Lumpur, Malaysia

K. H. Mo
Department of Engineering, School of Engineering and Technology, Sunway University, Bandar Sunway, Selangor Darul Ehsan, Malaysia

Q. Wang and A. Hosseini-Bandegharaei (eds.), *Environmental Functional Materials in Air Pollution*, Environmental Science and Engineering,
https://doi.org/10.1007/978-3-031-84571-0_3

emissions, is exploring greener alternatives (Supriya et al. 2023). The Global Cement and Concrete Association aims to produce carbon–neutral concrete by 2050, aligning with global climate targets. This involves improving energy efficiency, using cleaner fuels, capturing carbon emissions, and transitioning to alternative raw materials (Global Cement and Concrete Association 2023).

Biochar, a green alternative in construction, is a carbon-rich product from pyrolysis of waste biomass. Produced under limited oxygen at 500–700 °C, it has a porous structure and strong adsorption ability (Jagadeesh and Sundaram 2023). It's used for carbon sequestration and soil amendment and can reduce 870 kg of CO_2 equivalent per ton of biomass (Lu and Zong 2018). When incorporated into concrete, biochar recycles waste, captures carbon, and improves mechanical durability. It also exhibits low flammability, low thermal conductivity, and strong chemical stability within cement (Akinyemi and Adesina 2020). Its porous structure improves interaction with the cement matrix, enhancing the material's properties. Most of the research focuses on incorporating a small percentage of locally sourced Rice Husk Biochar (RHB) into mortar mixtures to examine its effects on mechanical properties, aiming to develop sustainable concrete solutions. Findings indicate that replacing 2–5% of traditional materials with biochar can enhance compressive strength by 3–4% (Barbhuiya et al. 2024).

In concrete, carbon sequestration occurs when unreacted cement and hydration products interact with CO_2 in the atmosphere to produce calcium carbonate. Natural carbonation typically takes place at the concrete surface and influences hardening, durability, and deterioration processes in cement-based materials. The efficiency of carbonation can be improved via accelerated carbonation process, which may help to offset emissions from production of cement. This method is conducted in an elevated CO_2 concentrations environment. The carbonation rate is affected by properties such as cement content, concrete strength, water-cement ratio, porosity, and the degree of water saturation in the concrete's pore structure (Barbhuiya et al. 2024). Incorporating biochar into concrete enhances carbonation through production of more hydration products. The pores in biochar act as nucleation sites, accelerating both the rate and depth of hydration. Additionally, biochar's high surface area improves diffusion of CO_2 by allowing more area for carbonation reactions through the strengthened cement matrix. Thus, Biochar has been found to positively impact the carbon capture performance of cement mixes (Senadheera et al. 2023).

The paper studies biochar's incorporation into cement mortar, assessing its impact on various properties and developing eco-friendly concretes. It also compares different curing conditions and quantifies carbonation depth and degree in the biochar cement composite.

Table 3.1 Oxide composition of OPC and RHB

Oxide	OPC	RHB
	(wt%)	
CaO	65.72	2.4
SiO_2	18.53	80.71
Fe_2O_3	4.47	0.62
Al_2O_3	3.59	–
MgO	0.93	–
K_2O	–	7.61
P_2O_5	–	6.58

3.2 Materials and Test Methods

RHB was substituted into cement mortar at 10, 20, 30, and 40% volume replacement of cement, namely RHB10, RHB20, RHB30, and RHB40. The Ordinary Portland Cement (OPC) utilized was an ASTM C150 Type I with 3.15 specific gravity (SG), 310 m2/kg of Blaine fineness, and 21.10 mm of median particle diameter (d_{50}). The water to binder ratio was 0.6 by mass. Specimens were cast into 50 mm cubes and hardened for 24 h, then subjected to water and carbonation curing. Water curing is at 23 ± 2 °C. The carbonation chamber is set at 20 °C, 80% humidity, and 20% CO_2. Compressive strength tests were conducted on 3d, 7d, and 28d of both curing methods. A sample from each replacement percentage was cut in half to check carbonation depth. Broken pieces from the strength test were ground and sieved for a TGA test. Table 3.1 presents the XRF results of major oxide composition for OPC and RHB.

RHB, analyzed by XRF, has high silica and meets ASTM-C618 class F, indicating potential pozzolanic properties. CHNS test shows RHB contains 33.4% carbon, 0.5% hydrogen, 9.1% nitrogen, 1.8% sulphur, and 55.1% oxygen. It has O/C and H/C molar ratios of 1.2 and 0.16 respectively, suggesting high aromaticity and stability (Muthukrishnan et al. 2019). FESEM reveals RHB's porous structure (Fig. 3.1).

3.3 Results and Discussion

The workability of mortar mixtures is assessed via flow table test (Table 3.2). As biochar weight percentage increases, flow diameter reduces. Fresh mortar flow is controlled at 220 ± 20 mm. 10% biochar flow is similar to control at 232.9 mm average. Biochar dosage increase from 20 to 40% reduces workability, requiring water correction. This is because of biochar's porous structure and its capacity to hold high volume of water (Yang and Wang 2021).

Table 3.3 presents the mix design and properties of the mortar. Biochar addition makes the cement mortar lighter with density decreases. Figure 3.2 presents the

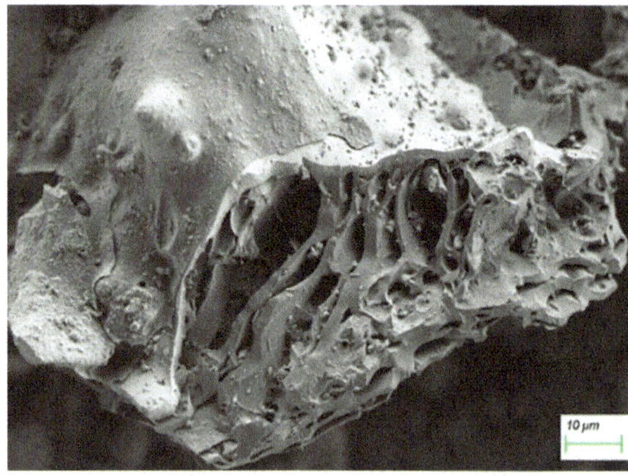

Fig. 3.1 Porous morphology of rice husk biochar

Table 3.2 Flow table test for mortar mix

	Control	RHB10	RHB20	RHB30	RHB40
Flow diameter (mm)	235.50 ± 3.00	230.39 ± 2.50	225.46 ± 3.20	218.07 ± 1.60	205.00 ± 3.70

compressive strength results at 3, 7, and 28 days. The overall compressive strength decreases with the addition of a high amount of porous biochar into the cement mortar. The more biochar replaces cement, the lower the compressive strength.

All specimens subjected to carbonation curing show an enhancement in compressive strength. For 3 days, there's an 82% strength gain for both the control and RHB10, followed by 67, 55, and 100% strength gain for RHB20, RHB30, and RHB40 respectively. For 7 days, the strength gain is 79, 66, 58, 53, and 69% for Control, RHB10, RH20, RHB30, and RHB40 respectively. The increase in compressive strength from 7 to 28 days of carbonation curing is < 15%, suggesting that the carbonation rate decreases after 7 days as most of the reactants have reacted with carbon dioxide and converted into calcium carbonate.

Biochar absorbs water during mixing, reducing free water in mortar pores and serving as an internal water source during hardening. Higher biochar replacement decreases compressive strength due to cement dilution and increased porosity,

Table 3.3 Mix design of RHB mortar in kg/m^3 unit

Mortar mix	Cement	Biochar	Water	Sand	Fresh mass	Fresh density
Control	500	0	300	1434.4	278	2227
RHB10	450	31.9	324.1	1434.4	277	2213
RHB20	400	63.9	348.3	1434.4	275	2200
RHB30	350	95.8	372.4	1434.4	269	2157
RHB40	300	127.8	397.6	1434.4	264	2116

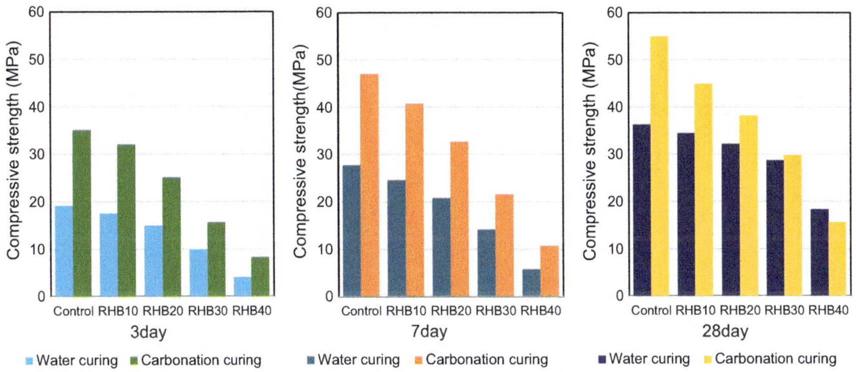

Fig. 3.2 Compressive strength of biochar cement mortar at 3 day, 7 day and 28 day for water curing and carbonation curing

making it suitable for lightweight, non-structural applications. A porous matrix enhances CO_2 uptake, and after carbonation curing, the compressive strength improves due to the formation of calcium carbonate.

Figure 3.3 shows that carbonation depth increases with more biochar, enhancing CO_2 uptake. The formation of calcium carbonate due to carbonation turns phenolphthalein indicator into colourless. Carbonation occurs from the surface towards the center of the biochar mortar. More carbonization is observed from 3 to 7 days, and RHB30 may be fully carbonized by the 7th day. Thus, the phenolphthalein test might not be necessary at the 28-day mark.

The TGA-DTG analysis identifies and quantifies hydration and carbonation products in the cement composite, which helps to quantifies the carbonation degree of mortar cubes. Figure 3.4 shows the TGA-DTG curve for control mortar in water and carbonation curing. Table 3.2 summarizes the carbonation degree for 3d and 7d carbonation curing. Figure 3.4 reveals four main regions and decomposition peaks representing weight losses when samples were heated to 950 °C. The first weight

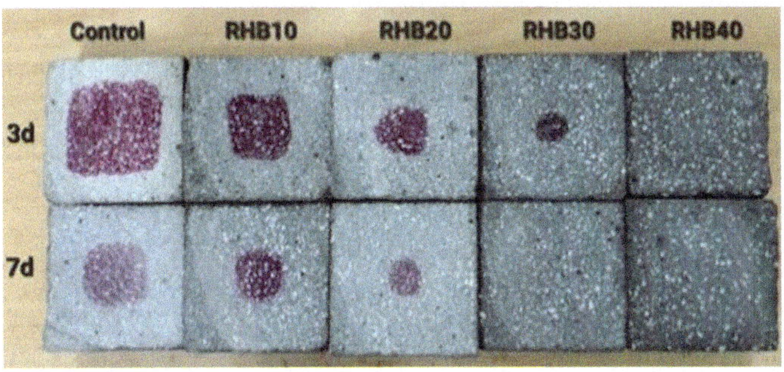

Fig. 3.3 Carbonation depth of different biochar replacement of mortar sprayed with phenolphthalein indicator

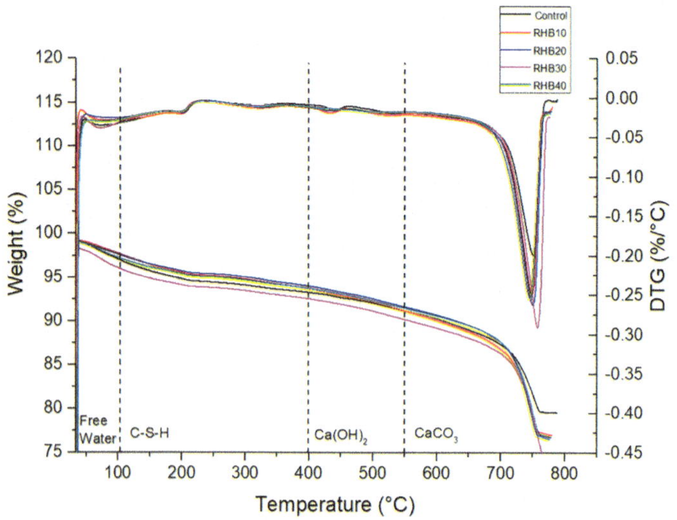

Fig. 3.4 The regions detected differences in TGA and DTG curves of plain cement paste subjected to carbonation curing

Table 3.4 Carbonation degree (%) for 3d and 7d carbonation curing

	Control	RHB10	RHB20	RHB30	RHB40
3 day	12.0	15.9	20.2	19.8	13.8
7 day	14.5	14.9	18.7	17.6	14.9

loss is before 105 °C, due to evaporation of free water or other volatiles. The second peak at 105–400 °C is due to dehydration of CSH gels. The third step between 400 and 550 °C is due to dissociation of calcium hydroxide product. The final stage between 550 °C and 800 °C indicates calcium carbonate formation in the mortar. For carbonation curing, the third section lacks a peak for calcium hydroxide as most of it was converted into calcium carbonate.

Based on TGA results Table 3.4 was calculated. It shows that the carbonation degree of biochar mortar increases with biochar dosage, peaking at RHB20. This trend is similar for both 3d and 7d, with the highest carbon uptake at 20.2% and 18.7% for RHB20. Adding biochar beyond RHB20 results in lower carbon uptake.

3.4 Conclusion

The addition of RHB to cement mortar impacts its pore structure, strength, and porosity. Biochar's water retention properties reduce flow and workability. As biochar content increases, compressive strength decreases, indicating its suitability for non-structural applications. However, accelerated carbonation curing can improve strength by up to 44% after 28 days of curing. TGA shows that more biochar improves carbonation and increases calcium carbonate formation, promoting carbon capture. The phenolphthalein test shows improved carbonation rate with RHB. Both carbonation depth and degree results prove that biochar facilitates carbonation curing of cement mortar. The optimum RHB dosage for maximum carbon uptake and carbonation degree is 20% volume replacement. Thus, adding biochar and accelerated carbonation curing offers a feasible method for carbon uptake and sequestration. In conclusion, incorporating biochar into concrete reduces waste and promotes carbon capture and storage through mineralization.

Acknowledgements The authors would like to thank the Ministry of Higher Education Malaysia for the Fundamental Research Grant Scheme (FRGS/1/2022/TK01/UM/02/2).

References

Akinyemi BA, Adesina A (2020) Recent advancements in the use of biochar for cementitious applications: a review. J Build Eng 32:101705

Barbhuiya S, Bhusan Das B, Kanavaris F (2024) Biochar-concrete: a comprehensive review of properties, production and sustainability. Case Stud Construct Mater 20:e02859

Global Cement and Concrete Association (2023) The GCCA 2050 cement and concrete industry roadmap for net zero concrete. https://gccassociation.org/concretefuture/. Accessed 7 Sept 2023

Jagadeesh N, Sundaram B (2023) Adsorption of pollutants from wastewater by biochar: a review. J Hazard Mater Adv 9:100226

Lu S, Zong Y (2018) Pore structure and environmental serves of biochars derived from different feedstocks and pyrolysis conditions. Environ Sci Pollut Res 25(30):30401–30409

Muthukrishnan S, Gupta S, Kua HW (2019) Application of rice husk biochar and thermally treated low silica rice husk ash to improve physical properties of cement mortar. Theor Appl Fract Mech 104:102376

Senadheera SS, Gupta S, Kua HW, Hou D, Kim S, Tsang DCW et al (2023) Application of biochar in concrete—a review. Cement Concr Compos 143:105204

Supriya, Chaudhury R, Sharma U, Thapliyal PC, Singh LP (2023) Low-CO_2 emission strategies to achieve net zero target in cement sector. J Clean Prod 417:137466

Yang X, Wang XY (2021) Strength and durability improvements of biochar-blended mortar or paste using accelerated carbonation curing. J CO_2 Utiliz 54:101766.

Chapter 4
Catalytic Degradation of Tetracycline by Core–Shell CoMn@NC Nanoparticles Via Enhanced Peroxymonosulfate Activation

Yuankang Bao, Xun Xu, and Dongbai Xie

Abstract This study synthesized core–shell structured CoMn@nitrogen-doped carbon (CoMn@NC) magnetic nanoparticles by annealing CoMn Prussian blue analog (PBA) precursors. The catalytic performance of the catalyst in activating peroxymonosulfate (PMS) to degrade tetracycline hydrochloride (TC) was investigated. Characterization of the catalyst using X-ray diffraction, and scanning electron microscopy revealed the presence of amorphous carbon, CoO, Co, and amorphous manganese species, with CoMn bimetallic particles encapsulated in nitrogen-carbon core–shell structures. At 25 °C and an initial pH of 6.2, with the addition of 50 mg/L catalyst and 200 mg/L PMS, complete degradation of 20 mg/L TC was achieved within 20 min.

Keywords Peroxymonosulfate activation · Bimetal catalyst · Wastewater treatment · Tetracycline degradation · Nitrogen-doped carbon

4.1 Introduction

In recent years, environmental and water pollution from antibiotic overuse has become severe (Milic et al. 2013). Primary impacts include increased microbial resistance, challenges to human health, degradation of water quality, and disruption of ecosystems (Lin et al. 2021; Huang et al. 2011). Tetracycline antibiotics are particularly concerning due to their stable structure and high hydrophilicity, making

Y. Bao · D. Xie (✉)
Weifang Key Lab of Advanced Light Materials Manufacturing and Forming, Weifang University of Science and Technology, Shouguang, China
e-mail: dbxie@aliyun.com

Y. Bao · X. Xu · D. Xie
Jiangxi Key Laboratory of Materials Surface Engineering, Jiangxi Science and Technology Normal University, Nanchang, China

Q. Wang and A. Hosseini-Bandegharaei (eds.), *Environmental Functional Materials in Air Pollution*, Environmental Science and Engineering,
https://doi.org/10.1007/978-3-031-84571-0_4

their degradation and removal challenging (Xu et al. 2021). Common methods for degrading tetracyclines include biodegradation, chemical oxidation, and photocatalysis, but these can suffer from inconsistent efficiency, toxic byproducts, and operational complexity (Wang and Zhuan 2020).

Advanced Oxidation Technologies (AOTs) like Fenton oxidation, UV oxidation, activated carbon oxidation, and titanium dioxide photocatalytic oxidation are effective for wastewater treatment and environmental remediation (Acero et al. 2018; Chowdury and Balasubramanian 2014). AOTs based on persulfate activation generate radicals that oxidize organic pollutants into harmless substances (Gode et al. 2019; Xiao et al. 2018). This technology is efficient, produces no toxic byproducts, and is environmentally friendly. Homogeneous Co^{2+} is efficient in PMS activation but causes secondary pollution and is not reusable. Heterogeneous Co-based catalysts improve ion leaching issues and offer superior performance (Tao et al. 2024). Co-M bimetallic oxides enhance efficiency and reduce ion leaching through synergistic effects, with manganese being a cost-effective activator (Ge et al. 2021; Hu et al. 2022). However, Co oxides and Co-M bimetallic oxides tend to agglomerate during reactions. Surface modification or coating with stabilizers can improve dispersion. Nitrogen doping in carbon materials improves pore structure, surface properties, and the stability of active species by enhancing interactions between nitrogen species and metals or metal oxides (Chen et al. 2020).

This study synthesized core–shell CoMn@NC bimetallic catalysts via co-precipitation and high-temperature annealing to activate peroxymonosulfate (PMS) for degrading tetracycline hydrochloride (TC). It systematically investigated the effects of initial TC concentration, catalyst dosage, pH, and temperature on degradation efficiency. Additionally, the study examined the potential interference from inorganic anions and natural organic matter in aquatic environments.

4.2 Experimental

4.2.1 Reagents and Instruments

Reagent: Cobalt nitrate hexahydrate ($Co(NO_3)_2 \cdot 6H_2O$), manganese nitrate solution ($Mn(NO_3)_2$ 50% wt%), cobalt potassium cyanide ($K_3[Co(CN)_6]$), polyvinylpyrrolidone (PVP), peroxymonosulfate (PMS), tetracycline hydrochloride (TC) and anhydrous ethanol are all analytically pure. Purchased from Shanghai McLean Biochemical Technology Co., LTD.

4.2.2 *Preparation of Core–Shell Comn@NC Bimetallic Catalyst*

In a standard synthesis procedure, 0.010 g of $K_3[Co(CN)_6]$, 0.10 g of PVP, 10.0 ml of water, and 10.0 ml of ethanol were combined and stirred continuously at room temperature for 1 h. Subsequently, 10.0 ml of a 3.3 wt% $Mn(NO_3)_2$ solution was added to the mixture, which was stirred for an additional four hours. Afterward, the mixture was centrifuged, and the resulting white precipitate was collected. This precipitate was washed several times with deionized water and ethanol, then air-dried. The resulting white powder was transferred to a porcelain boat and placed in a tube furnace, where it was annealed at 800 °C under an Ar gas atmosphere for 2 h, with a heating rate of 3 °C/min to produce CoMn@NC.

4.2.3 *TC Degradation Experiment*

Degradation experiments were carried out in 50 ml centrifuge tubes, with each experiment repeated three times. A specific amount of catalyst was initially added to 50 ml of an aqueous tetracycline (TC) solution and stirred for 30 min to allow for equilibrium adsorption of the catalyst onto the TC. Following this, PMS was introduced to commence the degradation process. At regular intervals, samples of the reaction solution were taken using a syringe and subsequently filtered through a 0.22 μm filter. The absorbance of the filtrate was recorded at the maximum absorption wavelength of TC (357 nm) using a UV–Vis spectrophotometer to assess the TC removal efficiency.

4.3 Results and Discussion

4.3.1 *Characterization of Catalysts*

The physical phase, crystallinity, and composition of CoMn@NC were analyzed using powder X-ray diffraction (XRD). In Fig. 4.1a, the broad peak near $2\theta = 26°$ indicates amorphous carbon. The diffractogram shows peaks corresponding to Co (JCPDS No. 75–0419) and CoO (JCPDS No. 89–4307) at 44.23°, 51.53°, 34.10°, and 57.22°, attributed to the (111) and (200) facets of Co and the (111) and (220) facets of CoO. No manganese oxides were detected, likely because they are amorphous. XPS tests were conducted to further investigate the chemical composition and electronic valence states. As shown in Fig. 4.1d, four elements C, Co, Mn, and O are present. Figure 4.1e and f display the high-resolution spectra before and after fitting of Co 2p and Mn 2p orbitals. The fitted peaks at 781.71 and 796.88 eV in Fig. 4.1e are Co^{2+} species, and the peaks at 779.89 and 795.28 eV are attributed to metallic Co,

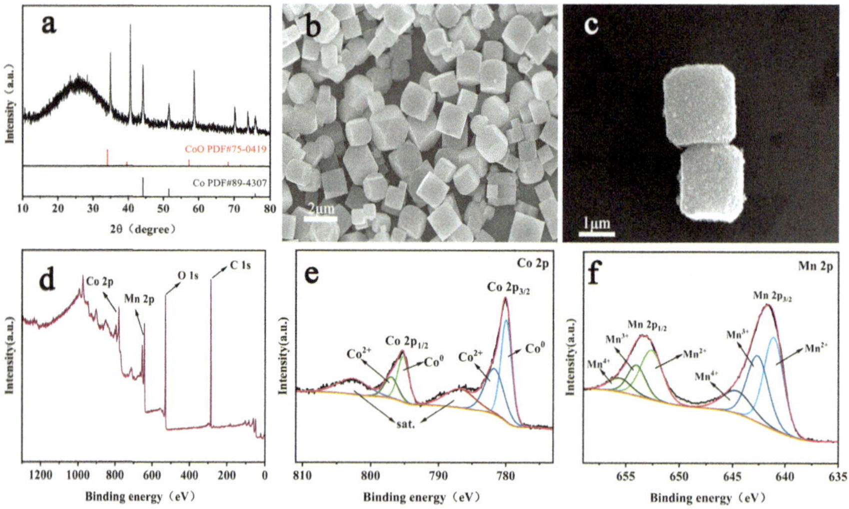

Fig. 4.1 **a** Powder XRD patterns of CoMn@NC samples. **b** SEM image of precursor CoMn PBAs. **c** SEM image of CoMn@NC. The XPS survey (**d**), Co 2p (**e**), Mn 2p (**f**)

consistent with the XRD results. The fitted peaks at 641.76 and 653.63 eV in Fig. 4.1f correspond to Mn^{2+} species, the peaks at 643.27 and 655.84 eV to Mn^{3+} species, and the peaks at 646.75 and 657.97 eV to Mn^{4+} species, confirming the presence of Mn species. SEM imaging (Fig. 4.1b) shows the precursor material as cubic with a smooth surface and an average size of about 1 μm. Figure 4.1c indicates that cobalt-manganese metal is uniformly encapsulated by nitrogen-carbon, preventing ionic dissolution and agglomeration.

4.3.2 Catalytic Degradation Performance of Catalysts

The catalytic performance of Co@NC and CoMn@NC in tetracycline (TC) degradation was evaluated by activating peroxymonosulfate (PMS) (Fig. 4.2a). Control experiments showed PMS alone degraded only 22.4% of TC, highlighting its ineffectiveness without a catalyst. The catalyst was equilibrated in the solution for 30 min before PMS addition. Both Co@NC and CoMn@NC removed about 30% of TC through adsorption. However, the CoMn@NC/PMS system demonstrated significantly higher TC degradation efficiency, exceeding 90% within 5 min, due to the synergistic effect of Co-Mn bimetallic oxides.

Figure 4.2b shows the impact of initial TC concentration on catalytic performance. As TC concentration increased, degradation efficiency decreased slightly. Concentrations of 10, 20, and 30 ppm were fully degraded within 20 min, while 50 ppm reached an 89.8% removal rate. In the first 5 min, the degradation rate dropped from

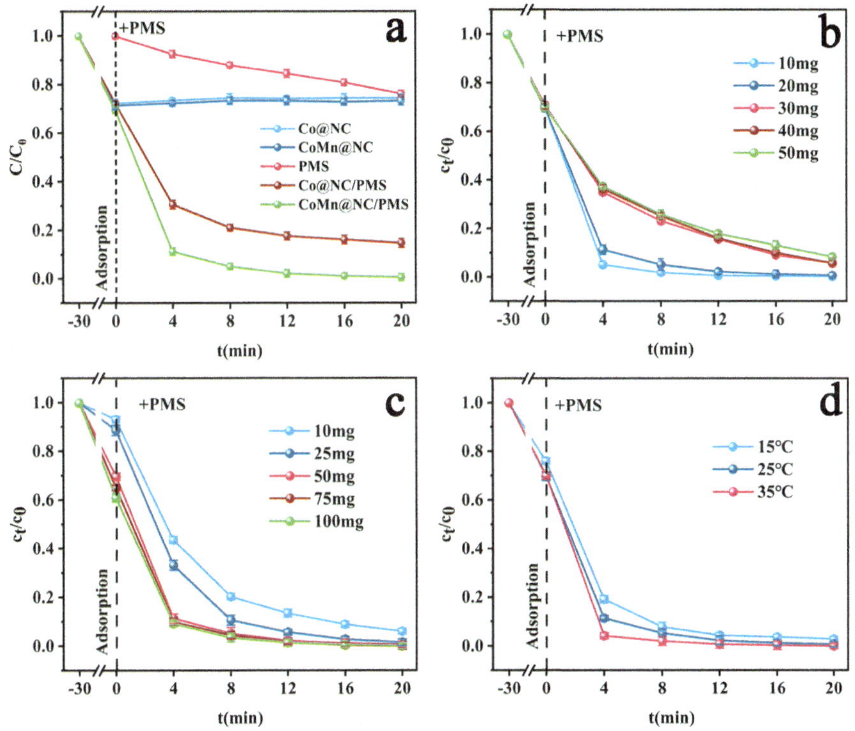

Fig. 4.2 **a** Degradation of TC in different reaction systems. Effects of initial concentration of TC (**b**), catalyst dosage (**c**), temperature (**d**) and initial pH (**e**) on TC degradation by CoMn@NC/PMS oxidation system. **f** Zeta-potential of CoMn@NC dispersion solution at various pH values

90 to 79% as the concentration increased. Figure 4.2c shows that increasing the catalyst dosage progressively raised the TC degradation rate, achieving 100% within 20 min at 50 mg/L. Higher dosages did not significantly enhance degradation, only increasing pollutant adsorption. Thus, 50 mg/L was chosen for subsequent studies to balance practical application and cost. Since the activation energy is a function of temperature, the rate of pollutant degradation is strongly influenced by the reaction temperature. Figure 4.2d shows that increasing the temperature from 15 to 35 °C significantly enhanced both adsorption and degradation efficiencies, indicating that the catalyst performs better in high-temperature wastewater conditions.

The degradation performance of the catalysts on TC was examined at pH values of 3.5, 5.1, 6.4 (initial), 7.5, 9.5, and 11, due to the sensitivity of cobalt-based catalysts to pH. Figure 4.2e shows that the catalysts maintained high efficiency between pH 3.5 and 9.5, with a TC removal rate of 93.2% at pH 3.5, indicating good performance across a broad pH range. As pH increased, TC adsorption on CoMn@NC was enhanced, attributed to changes in zeta potential (Fig. 4.2f) and the charge of TC. The negative zeta potential of CoMn@NC decreased with increasing pH, improving its electrical properties, reducing agglomeration, and providing more active sites. Additionally, higher pH caused TC molecules to become positively charged, enhancing interaction with CoMn@NC and improving adsorption capacity.

4.3.3 Mechanism of TC Degradation by CoMn@NC /PMS Oxidation System

The reactive species generated during TC degradation via activated PMS were investigated through classical radical scavenging experiments. Methanol (MeOH) was employed to scavenge hydroxyl and sulfate radicals, tert-butanol (TBA) for hydroxyl radicals, p-benzoquinone (p-BQ) for superoxide radicals, and L-histidine (L-His) for singlet oxygen. As depicted in the Fig. 4.3a, the addition of MeOH and TBA to the CoMn@NC/PMS system resulted in a decrease in TC removal by 35 and 4.3%, respectively, indicating that $SO_4^{\cdot-}$ was the primary active species responsible for TC degradation, with minimal presence of •OH radicals. The introduction of 1 mM L-His led to a 58% decrease in TC degradation, indicating the important role of $^1O_2$ in TC degradation, which was attributed to the unique metal-nitrogen coordination and appropriate defects in CoMn metal-nitrogen-carbon materials, which could effectively activate PMS to produce $^1O_2$. Moreover, the presence of p-benzoquinone caused only a 2.8% reduction in TC removal rate, suggesting minimal contribution of $O_2^{\cdot-}$ in TC degradation. To confirm the generation of reactive oxygen species in the oxidation systems, in situ EPR experiments were performed using DMPO and TEMP as spin trapping agents. As shown in Fig. 4.3b–d, no radical signals were detected in the EPR spectra when only PMS was added to the reaction solution without a catalyst at 0 min, indicating that PMS alone did not produce radicals. However, when PMS was added simultaneously with CoMn@NC, clear DMPO-$SO_4^{\cdot-}$, DMPO-OH, and TEMP-$1O_2$ adduct signals appeared in the EPR spectra, confirming the generation of $SO_4^{\cdot-}$, ·OH, and $1O_2$, respectively.

4.3.4 Reusability of CoMn@NC for TC Degradation

To further investigate the effectiveness of CoMn@NC, experiments were conducted under various conditions. The CoMn@NC/PMS system showed strong resistance to inorganic anions and natural organic compounds such as Cl^-, $CO_3{}^{2-}$, $NO_3{}^-$, and humic acid (HA) (Fig. 4.4a). Enhanced degradation was observed with Cl^- due to the formation of HOCl, which accelerates catalytic activity. $PO_4{}^{3-}$ slightly inhibited the process by occupying active sites on the catalyst surface. Additionally, the system demonstrated excellent resistance to interference from complex matrices like lake water, groundwater, and tap water, achieving TC removal efficiencies above 95% (Fig. 4.4b), confirming its practical utility.

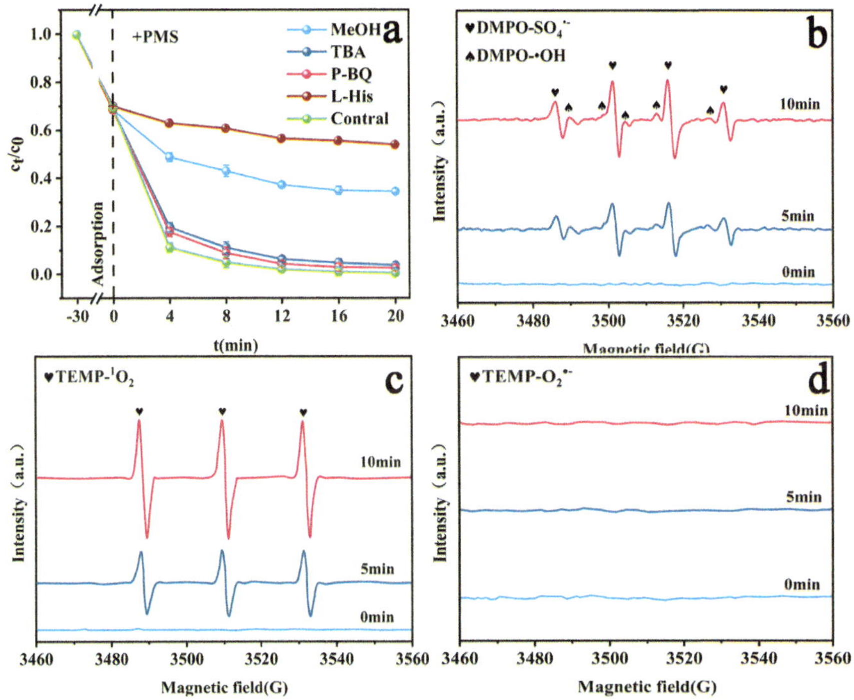

Fig. 4.3 **a** Effect of scavengers on TC degradation by CoMn@NC/PMS oxidation system. EPR trapping signals of DMPO-$SO_4^{\cdot-}$ and DMPO-·OH (**b**), and TEMP-$^1O_2$ (**c**), and DMPO-$O_2^{\cdot-}$ (**d**)

We performed four 20-min reaction cycles to maintain consistent conditions (Fig. 4.4c). After each cycle, the catalyst was recovered, washed with anhydrous ethanol and ultrapure water, and dried at 60 °C. Although the TC removal rate slightly decreased with increasing cycles, likely due to intermediate products adsorbed on the catalyst surface, the catalyst still achieved 92.04% TC degradation after the fifth reuse. Additionally, the CoMn@NC/PMS system exhibited a total organic carbon (TOC) removal rate of 97.3% in 20 min (Fig. 4.4d). These results indicate that the catalyst has excellent stability, is recyclable for multiple cycles, and is highly effective for treating organic pollutants in wastewater.

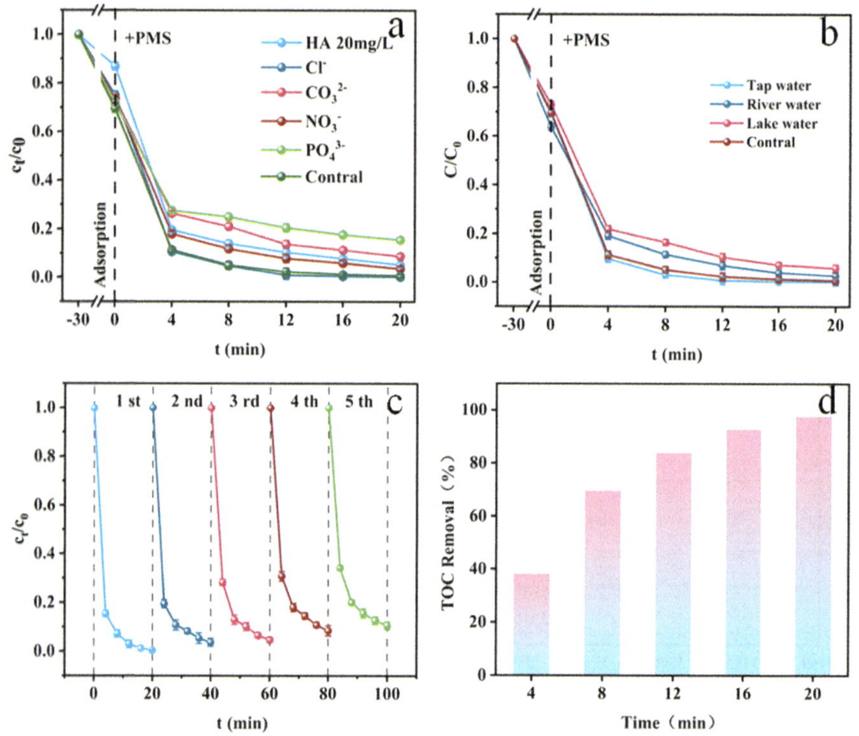

Fig. 4.4 The feasibility of CoMn@NC /PMS for removing TC in **a** different concentrations of HA and inorganic anions, **b** actual water bodies. **c** Cyclic tests of CoMn@NC/PMS for TC removal. **d** TOC removal efficiency of TC

4.4 Conclusion

This study synthesized CoMn@NC bimetallic catalysts with a core–shell structure via co-precipitation and high-temperature annealing, effectively activating PMS to degrade TC pollutants. Characterization by XRD and SEM revealed the catalyst consisted of amorphous carbon, CoO, Co, and amorphous manganese species, with nitrogen-doped carbon coating the CoMn particles. Under room temperature, initial pH 6.2, TC 20 mg/L, PMS 0.20 g/L, and catalyst dosage 0.05 g/L, the TC removal rate reached 100% in 20 min. The catalyst showed effective PMS activation from 15 to 35 °C and pH 3.5–9.5. Higher temperatures and alkaline pH accelerated TC degradation. Cl^- promoted TC degradation, NO_3^- and HA had minimal impact, and PO_4^{3-} slightly inhibited degradation, but the TC removal rate was 83.68% in 20 min. The system's performance was minimally affected by the water environment, suitable for real wastewater. After five cycles, the catalyst maintained a 91.34% TC degradation rate, demonstrating excellent stability. This study highlights the potential of nitrogen-doped carbon-coated bimetallic catalysts for PMS activation in antibiotic wastewater treatment.

Acknowledgements The authors would like to thank the financial supports of the Weifang University of Science and Technology.

References

Acero JL, Javier Benitez F, Real FJ, Rodriguez E (2018) Separ Purif Technol 201:41–50
Chen Z, Yang W, Wu Y, Zhang C, Luo J, Chen C, Li Y (2020) Nano Res 13:3075–3081
Chowdhury S, Balasubramanian R (2014) Adv Colloid Interf Sci 204:35–56
Ge M, Hu Z, Wei J, He Q, He Z (2021) J Alloys Compd 888
Gode JN, Souza DH, Trevisan V, Skoronski E (2019) J Environ Chem Eng 7
Hu Y, Chen D, Wang S, Zhang R, Wang Y, Liu M (2022) Separ Purif Technol 280
Huang J-J, Hu H-Y, Tang F, Li Y, Lu S-Q, Lu Y (2011) Water Res 45:2775–2781
Lin Z, Yuan T, Zhou L, Cheng S, Qu X, Lu P, Feng Q (2021) Environ Geochem Health 43:1741–1758
Milic N, Milanovic M, Letic NG, Sekulic MT, Radonic J, Mihajlovic I, Miloradov MV (2013) Int J Environ Health Res 23:296–310
Tao L, Zhang J, Yu F (2024) New J Chem 48:1757–1770
Wang J, Zhuan R (2020) Sci Total Environ 701:135023
Xiao R, Luo Z, Wei Z, Luo S, Spinney R, Yang W, Dionysiou DD (2018) Curr Opin Chem Eng 19:51–58
Xu L, Zhang H, Xiong P, Zhu Q, Liao C, Jiang G (2021) Sci Total Environ 753:141975

Chapter 5
Kinetic Analysis of N-Desorption in Chemical Looping Ammonia Synthesis

Zhongyuan Liu, Qingbo Yu, and Jiatai Zhao

Abstract This study examines how varying reaction temperatures influence the N-desorption process, along with the development of a kinetic model for the reaction. The results indicate that higher temperatures enhance the N-desorption reaction, and that the shrinking core model provides the best fit for describing the reaction kinetics. The activation energy associated with the N-desorption process is 330.52 kJ/mol.

Keywords Chemical looping ammonia synthesis · N-desorption kinetics · Reaction modeling

5.1 Introduction

Low-carbon or carbon-free energy sources are critical to addressing the global energy crisis and reducing emissions (Liu 2024). Ammonia (NH_3) represents a promising carbon-free energy carrier with high energy density and is also advantageous due to its lower cost and easier handling for transport and storage compared to hydrogen (H_2). Currently, over 90% of global ammonia production relies on the Haber–Bosch process. However, this traditional approach has notable limitations, including high CO_2 emissions, substantial energy consumption, and relatively low efficiency. Consequently, there is a pressing need for an alternative, low-carbon, environmentally friendly, and efficient ammonia synthesis method (Liu et al. 2023).

Chemical looping technology, as a promising solution, divides the overall reaction into multiple steps, using intermediate products as a circulating medium. This approach effectively circumvents the thermodynamic and kinetic limitations inherent in traditional ammonia synthesis processes. By structuring reactions in this way, it not only increases reaction rates but also significantly enhances the final conversion efficiency. Chemical looping technology has already found widespread applications in

Z. Liu · Q. Yu (✉) · J. Zhao
School of Metallurgy, Northeastern University, Shenyang, Liaoning, People's Republic of China
e-mail: yuqb@mail.neu.edu.cn

Q. Wang and A. Hosseini-Bandegharaei (eds.), *Environmental Functional Materials in Air Pollution*, Environmental Science and Engineering,
https://doi.org/10.1007/978-3-031-84571-0_5

various fields, including chemical looping oxygen production, hydrogen production, and combustion.

Based on the concept of chemical looping, the Chemical Looping Ammonia Generation (CLAG) process was proposed. CLAG divides the ammonia synthesis process into two primary steps: nitrogen absorption and nitrogen desorption. In the nitrogen absorption step, high temperatures break the N≡N triple bond, allowing nitrogen to be adsorbed and activated. In the subsequent nitrogen desorption step, the adsorbed nitrogen reacts with hydrogen to produce ammonia. Depending on the hydrogen source for the desorption reaction, chemical looping ammonia synthesis can be further divided into hydrogen-based chemical looping ammonia generation (H_2-CLAG) and water-based chemical looping ammonia generation (H_2O-CLAG). This stepwise reaction design markedly improves the efficiency and selectivity of ammonia synthesis, providing a new direction for the development of more efficient, low-carbon ammonia production technologies.

In H_2-CLAG, metal nitrides capture nitrogen atoms from nitrogen gas in the nitrogen absorption reactor. These metal nitrides then enter the nitrogen desorption reactor, where they react with hydrogen to produce ammonia, releasing nitrogen atoms in the process. These atoms are then cycled back to the nitrogen absorption reactor, creating a continuous loop. Michalsky et al. (2015) used thermodynamic screening to select combinations of nitrogen-poor and nitrogen-rich transition metal nitrides as nitrogen carriers. Their results showed that manganese nitrides, with multiple valence states, offer thermodynamically feasible conversion pathways. Laassiri et al. (2018) experimentally investigated the application potential of manganese nitrides in H_2-CLAG, synthesizing A-Mn-N (A = Fe, Co, K, Li) materials at low temperatures using $NaNH_2$ as a reactive nitrogen source, and subsequently assessing their performance. The results indicated that lithium doping significantly improved the nitrogen absorption capability of manganese nitride compared to pure Mn_3N_2; however, the lithium-doped Mn-N compound showed high stability after nitrogen absorption, limiting its activity in the nitrogen desorption reaction. Wang et al. studied the nitrogen-carrying capabilities of Fe and Mn metal nitrides and successfully achieved ammonia synthesis via chemical looping at 700–900 °C. Their results showed that nitrogen absorption activity ranked as Mn > Mn-Fe bimetallic > Fe, while nitrogen desorption activity ranked as Fe > Mn-Fe bimetallic > Mn. The Mn-Fe bimetallic nitrogen carrier effectively compensated for the low nitrogen absorption performance of Fe-based carriers and the low nitrogen desorption performance of Mn-based carriers (Wang et al. 2023). Wang et al. (2020) further investigated Al_2O_3 as a support material for Mn-Fe bimetallic nitrogen carriers and found that with a low Al_2O_3 content (≤ 5 wt%) and moderate reaction temperature (800 °C), sintering issues were alleviated, and ammonia yield was significantly increased. Gao et al. (2018) proposed a nitrogen carrier pair comprising alkaline earth metal hydrides and imide compounds ($AH_x/A_{2/x}NH$, where A = Li, Na, K, Mg, Ca, Ba). Thermodynamic calculations indicated that the BaH_2-BaNH and LiH-Li_2NH systems have moderate thermodynamic properties, making them potential nitrogen carrier pairs.

The H_2O-CLAG method, introduced by Galvez et al. (2007a, b), allows for ammonia production at atmospheric pressure by dividing the synthesis process into two key steps: nitrogen absorption and nitrogen desorption. During the nitrogen absorption (N-absorption) step, nitrogen gas is absorbed by a metal or metal nitride carrier. This carrier is then transferred to the nitrogen desorption (N-desorption) step, where, in the presence of hydrogen, the absorbed nitrogen is released and reacts to form ammonia. This division of the synthesis process into distinct absorption and desorption steps helps optimize reaction conditions and reduce energy demands, offering a promising pathway for more sustainable ammonia production.

N-absorption reaction:

$$N_2 + 3C + Al_2O_3 \rightarrow 2AlN + 3CO \quad \Delta H^0_{25\,^\circ C} = 708.1\ kJ/mol \tag{5.1}$$

N-desorption reaction:

$$2AlN + 3H_2O \rightarrow Al_2O_3 + 2NH_3 \quad \Delta H^0_{25\,^\circ C} = -274.1\ \ kJ/mol \tag{5.2}$$

Researchers have explored various aspects of CLAG to optimize its efficiency. Fang et al. (2022) investigated how different types of coal-derived coke affect the N-absorption step in CLAG, examining the influence of reaction conditions. Similarly, Weng et al. (2021) analyzed the economic viability and energy requirements of using biomass as a feedstock in the CLAG process. Despite these studies, research on the reaction kinetics of CLAG remains incomplete. The kinetics of the N-desorption reaction, in particular, are essential for guiding the precise control and scale-up of CLAG.

In this paper, the isothermal tests investigated the kinetic modeling of the N-desorption reaction. Initially, the impact of reaction temperature on the N-desorption process was examined, resulting in transient reaction curves at various temperatures. Then, the reaction curves were fitted with different kinetic models to obtain the most compatible kinetic model. Ultimately, a kinetic framework was developed for the N-desorption reaction by obtaining the frequency factor and activation energy via linear regression analysis. The theoretical results presented in this study offer guidance for the industrialization of CLAG.

5.2 Experiments

5.2.1 Experiment Process

Figure 5.1 shows the experimental system. A total of 0.82 g of AlN was placed in a crucible, which was then suspended from an electronic balance and positioned at the center of a tube furnace. Throughout the experiment, the temperature of the tube furnace was raised slowly up to the set temperature within an N_2 environment at a ramp rate equal to 10 °C/min. After reaching the target temperature, the atmosphere was changed to steam with a concentration of 80 vol% and a flow rate set to 500 ml/

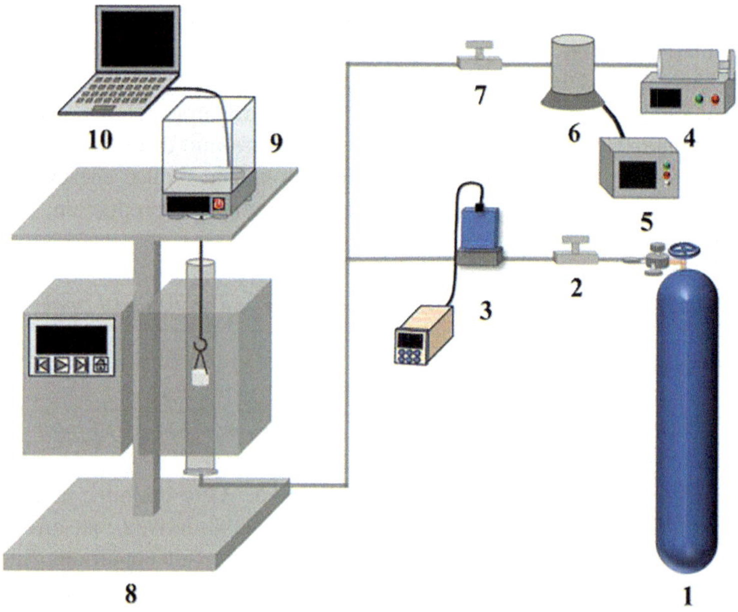

Fig. 5.1 Thermogravimetric experimental system diagram (1: N_2 cylinder; 2: valve; 3: Flowmeter; 4: syringe pump; 5: temperature controller; 6: steam generator; 7: valve; 8: tube furnace; 9: electronic balance; 10: computer)

min. The reaction was performed for 1 h, during which the changes in sample mass were recorded using a computer. The AlN conversion was determined according to the relationship given below:

$$\alpha_{\mathrm{A}} = \frac{m_t / M_{\Delta m}}{m_{\mathrm{AlN}} / 2M_{\mathrm{AlN}}} \times 100\% \tag{5.3}$$

where α_{A} is the AlN conversion, %; m_t denotes the change in mass for the sample, g; m_{AlN} represents the AlN mass present within the sample, which is 0.82 g; and M_{AlN} indicates the molar mass of AlN, 41 g/mol.

5.2.2 *Kinetic Calculation*

The kinetic model applied to the N-desorption reaction incorporates various reaction orders and diffusion models, including Am (m = 1, 2, 3, 4), Rm (m = 1/2, 1/3, 1/4, 2, 3), Dm (m = 1, 2, …, 7, 8), and Cm (m = 1, 2). The reaction rate is defined by Eq. (5.4), and through transformation, Eq. (5.5) can be derived. Using linear regression analysis, the most suitable kinetic model for the N-desorption reaction can be identified. Additionally, the activation energy and frequency factor are calculated using Eq. (5.6), providing insights into the reaction's temperature dependence and kinetic parameters. This modeling approach aids in understanding the N-desorption process's dynamics and optimizing reaction conditions for CLAG applications.

$$\frac{d\alpha_A}{f(\alpha_A)} = k(T)dT \tag{5.4}$$

$$\int_0^{\alpha_1} \frac{d\alpha}{f(\alpha)} = F(\alpha) = \int_0^{t_1} k(T)dt = K(T) \cdot t \tag{5.5}$$

$$\ln k(T) = -\frac{E}{RT} + \ln A \tag{5.6}$$

5.3 Results and Discussion

5.3.1 Impact of Reaction Temperatures

Figure 5.2 illustrates the impact of reaction temperature on both the conversion rate and reaction rate, providing valuable insight into the thermal dynamics governing the process. Figure 5.2a depicts the variation in conversion rate (α_A, %) over time at temperatures ranging from 975 °C to 1075 °C. The results demonstrate that conversion rates are significantly influenced by temperature, with a clear trend of increased conversion at higher temperatures. At 975 °C, the reaction proceeds very slowly, with minimal conversion even after 60 min, suggesting that the thermal energy provided is insufficient to drive the reaction forward effectively. In contrast, as the temperature increases to 1075 °C, the conversion rate accelerates, reaching values near 0.8 within the same time frame, which indicates a substantial enhancement in reaction efficiency.

This observation is consistent with the endothermic nature of the reaction, where higher temperatures provide the necessary energy input to overcome activation barriers, thus facilitating the desorption and subsequent conversion processes. As the reaction is endothermic, an increase in temperature supplies additional energy, promoting the breakdown of chemical bonds and enabling greater interaction among reactive species. Consequently, temperatures above 1000 °C appear to provide the optimal conditions required to maximize conversion efficiency over the observed timeframe.

Figure 5.2b further elucidates the effect of temperature by illustrating the variation in reaction rate (k) over time under identical temperature conditions. A distinct trend is observed in which the reaction rate initially rises sharply, reaches a peak, and subsequently declines. The data indicate that as temperature increases, the maximum reaction rate also increases, and the peak occurs earlier in the reaction, particularly at 1075 °C. This pattern implies that higher temperatures not only accelerate the initial

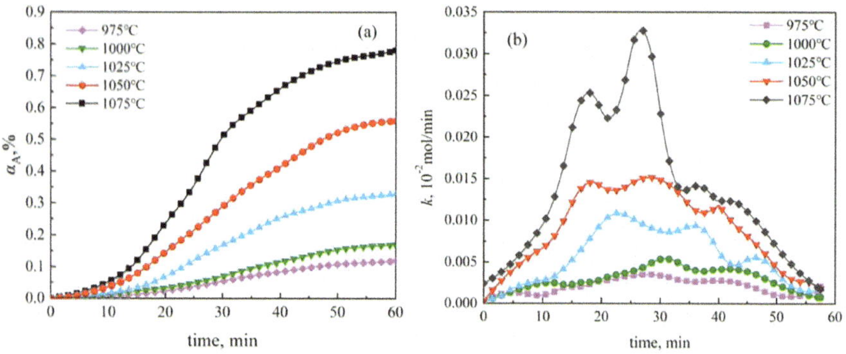

Fig. 5.2 Effect of reaction temperatures **a** conversion rate, **b** reaction rate

reaction kinetics but also lead to a more rapid consumption of reactants, thereby resulting in a subsequent decline in rate. The sharper and higher peaks at elevated temperatures highlight the increased reactivity but also point towards potential limitations, such as reactant depletion or changes in the reaction mechanism that may reduce efficiency over time.

Overall, these findings suggest that the reaction benefits significantly from elevated temperatures, with higher conversion rates and reaction speeds being achieved. However, the decline in reaction rate after reaching a peak indicates that while increasing temperature enhances the reaction's initial progression, careful optimization is required to balance the temperature to sustain a steady reaction rate and avoid rapid depletion of reactants. These insights are crucial for optimizing operational conditions in industrial applications where controlled and efficient conversion is desired.

5.3.2 *Kinetic Model Selection of N-desorption Reaction*

Figure 5.3 shows the $G(\alpha_A) \sim t$ linear fitting curves at various reaction temperatures, specifically: (a) 975 °C, (b) 1000 °C, (c) 1025 °C, (d) 1050 °C, and (e) 1075 °C. Linear fitting analysis is an essential approach for understanding the kinetic behavior of the N-desorption reaction at different temperatures. Each curve illustrates the progression of conversion over time, highlighting the effect of temperature on the reaction dynamics. The results indicate that as the temperature increases, the linearity of the fitting curves improves, signifying a more predictable kinetic pattern at elevated temperatures. By evaluating the average R^2 values obtained from the fitting curves at different temperatures, the R3 model was identified as the most appropriate to describe the kinetics of the N-desorption reaction. The R3 model, characterized by its unique mathematical framework for describing complex reaction mechanisms, provided the best correlation among the tested models, reflecting a strong relationship between experimental data and theoretical predictions. The selection of the R3 model for this reaction is particularly important as it underscores the role of diffusion and the complex nature of the desorption process in determining the overall reaction rate.

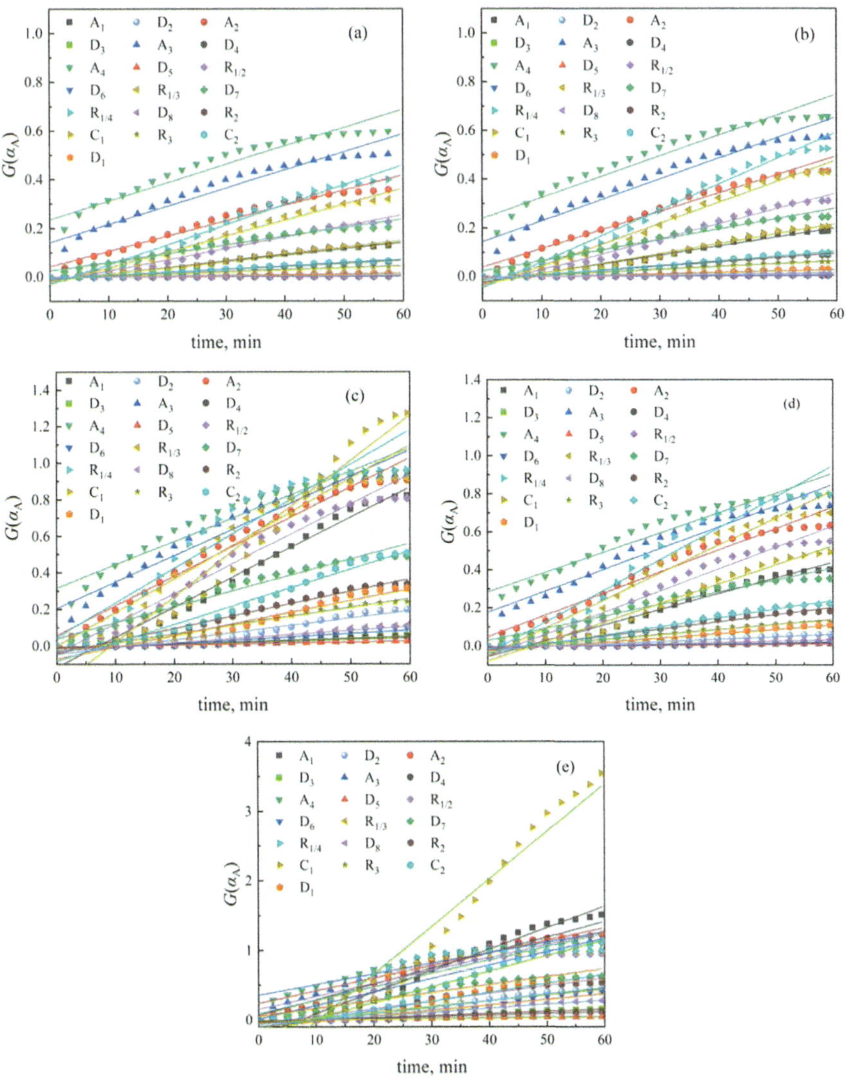

Fig. 5.3 G(αA) ~ t linear fitting curves at various reaction temperatures **a** 975 °C, **b** 1000 °C, **c** 1025 °C, **d** 1050 °C, **e** 1075 °C

5.3.3 Establishment of Kinetic Model

Figure 5.4 presents the linearly fitted curves for ln(k(T)) versus 1/T, which were used to establish the kinetic model for the N-desorption reaction. Table 5.1 shows the reaction rate coefficient and the activation energy. The activation energy (Ea) for the N-desorption process was determined from the slope and intercept of the fitted line. The calculated activation energy was found to be 330.52 kJ/mol, suggesting a high energy requirement for the reaction to proceed effectively. This value is indicative of the strong chemical bonds involved in the nitrogen desorption process, which require substantial energy input to overcome. Equation 5.7 represents the kinetic model derived from the experimental data and provides a quantitative means to predict the rate of reaction under various temperature conditions. Such a model is crucial for scaling up the reaction from laboratory to industrial settings, allowing for optimization of operational parameters to achieve desired conversion rates.

$$\frac{d\alpha_A}{dt} = 5.154 \times 10^{10}[\exp(-\frac{24.666}{T})] \cdot 3(1-\alpha_A)^{2/3} \tag{5.7}$$

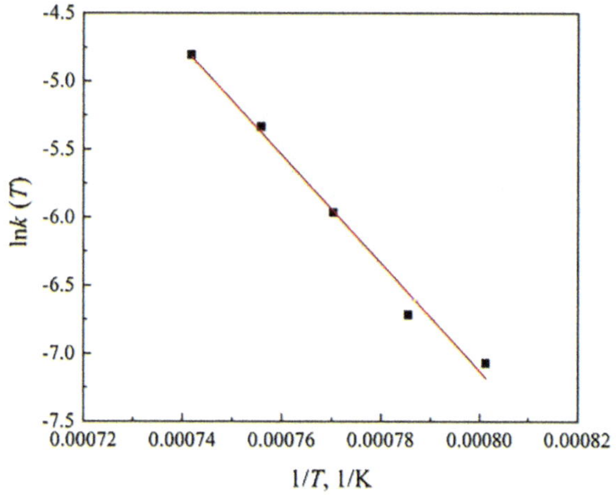

Fig. 5.4 Linearly fit curves for ln*k*(*T*) versus 1/*T*

Table 5.1 The reaction rate coefficient and the activation energy

Intercept	Slope	E_a (kJ/mol)	k_0 (min^{-1})	R^2
24.66	-39,755.61	330.52	5.17E10	0.989

5.4 Conclusions

In this paper, the N-desorption reaction was examined at different temperatures. It was found that higher temperatures positively influence the N-desorption process, leading to increased reaction rates and conversion efficiencies. The reaction kinetics were best represented by the R3 model, which provided the highest average R^2 values, indicating a strong correlation between the model and experimental data. The activation energy for the N-desorption reaction was calculated to be 330.52 kJ/mol, reflecting the significant energy required to drive the desorption process. These findings contribute to a better understanding of the thermal behavior and kinetics of the N-desorption reaction and provide a foundation for optimizing operational conditions in industrial applications to achieve efficient and controlled conversion processes.

References

Fang J et al (2022) Utilization of carbon-based energy as raw material instead of fuel with low CO_2 emissions: energy analyses and process integration of chemical looping ammonia generation. Appl Energy 312:118809

Galvez ME, Halmann M, Steinfeld A (2007b) Ammonia production via a two-step Al_2O_3/AlN thermochemical cycle. 1. Thermodynamic, environmental, and economic analyses. Indus Eng Chem Res 46(7):2042–2046

Galvez ME et al (2007a) Ammonia production via a two-step Al2O3/AlN thermochemical cycle. 2. Kinetic analysis. Indus Eng Chem Res 46(7):2047–2053

Gao W et al (2018) Production of ammonia via a chemical looping process based on metal imides as nitrogen carriers. Nat Energy 3(12):1067–1075

Laassiri S et al (2018) The potential of manganese nitride based materials as nitrogen transfer reagents for nitrogen chemical looping. Appl Catal B Environ 223:60–66

Liu Z et al (2023) Experimental investigation on centrifugal granulation of molten yellow phosphorus slag. Chem Eng Res Des 197:548–557

Liu Z et al (2024) Effect of pyrolysis parameters on the biochar reactivity in the N-absorption reaction of chemical looping ammonia generation. Energy 310:133321

Michalsky R et al (2015) Chemical looping of metal nitride catalysts: low-pressure ammonia synthesis for energy storage. Chem Sci 6(7):3965–3974

Wang B et al (2020) N-sorption capability of Al_2O_3-supported Mn-/Fe-based nitrogen carriers during chemical looping ammonia synthesis technology. Energy Fuels 34(8):10247–10255

Wang B et al (2023) Chemical looping ammonia synthesis at atmospheric pressure benefiting from synergistic effect of Mn- and Fe-based nitrogen carriers. Int J Hydrogen Energy 48(7):2705–2717

Weng Q et al (2021) Ammonia production from biomass via a chemical looping-based hybrid system. J Clean Prod 289:125479

Chapter 6
Investigation on the Condensation Effect of Particulate Matter Under the Action of Sound Waves in Different Working Conditions

Tianli Jiang, Xiang Gao, Donglin Yang, Lintao Qi, and Jin Wang

Abstract The prevention and reduction of PM2.5 and other pollution particles is a major challenge environmental protection. The existing emission reduction methods are complex and costly, and secondary pollution needs to be dealt with. In this study, acoustic agglomeration technology is proposed to promote the accumulation of particles through acoustic waves. This method is environmentally friendly and efficient, and shows a good application prospect. Sound wave condensation technology promotes the relative movement of pollution particles through the strong sound field, and the fluid medium will produce periodic changes when the sound wave propagates. The difference in response of particles of different scales to this change leads to different motion speeds, and the collision polymerization probability between particles increases. After multiple collisions, the fine particles condense into larger particles, which settle under the action of gravity, effectively reducing the pollution particles in the air. The team carried out a simulation study, using the control variable method to explore the agglomeration efficiency of sound waves on particles under different working conditions. By applying acoustic waves with the same sound pressure level and different frequencies to particles, it is concluded that the effect of acoustic wave frequency on the agglomeration effect of particles is not a simple linear relationship, but there is an optimal frequency, which is between 600–1200 Hz, and the agglomeration effect of particles reaches the best in this frequency range. Then, the sound wave is controlled within the optimal frequency range, and the sound wave with different sound pressure levels is applied to the particles. The experimental results show that the sound pressure level is positively correlated with the condensation efficiency of the pollution particles, but its increase is not linear. When the sound pressure level is lower than 140 dB, the increase in condensation efficiency of pollution particles is not significant for each increase of 10 dB. Finally, the team designed the relevant device for experimental demonstration, which proved the correctness of the simulation conclusion and the reality of the method.

T. Jiang (✉) · X. Gao · D. Yang · L. Qi · J. Wang
National University of Defense Technology University, Changsha, China
e-mail: 15616213665@163.com

Q. Wang and A. Hosseini-Bandegharaei (eds.), *Environmental Functional Materials in Air Pollution*, Environmental Science and Engineering,
https://doi.org/10.1007/978-3-031-84571-0_6

Keywords Sound waves · Acoustic wave condensation · Simulation

6.1 Introduction

Air pollution has emerged as one of the most pressing environmental challenges of the twenty-first century, with particulate matter (PM) being a significant component of this crisis. Particulate matter, which includes a complex mixture of solid particles and liquid droplets suspended in the air, can vary in size, composition, and origin. The most concerning fractions are PM2.5 (particles with a diameter of 2.5 μm or smaller) and PM10 (particles with a diameter of 10 μm or smaller), due to their ability to penetrate deep into the respiratory system and cause a myriad of health problems. The World Health Organization (WHO) has estimated that air pollution is responsible for approximately 7 million premature deaths worldwide each year, with a substantial portion attributable to PM exposure. As such, effective strategies for particulate matter reduction are not only essential for environmental sustainability but also for public health protection.

The sources of particulate matter are multifaceted and can be broadly categorized into natural and anthropogenic origins. Natural sources include volcanic eruptions, wildfires, and dust storms, while anthropogenic sources primarily stem from industrial processes, vehicular emissions, construction activities, and the burning of fossil fuels for energy. The increasing urbanization and industrialization, particularly in developing countries, have exacerbated the problem, leading to deteriorating air quality and heightened health risks for urban populations. The complexity of PM composition, which can include heavy metals, organic compounds, and biological materials, further complicates the assessment of its health impacts and the development of effective mitigation strategies. In recent years, global efforts to address air quality issues have gained momentum. In 2018, Zhang Jing (Yang et al. 2023) conducted a systematic study on the emission reduction methods of particulate matter generated by diesel vehicles, mainly investigating the change law of particulate matter emission reduction performance of three kinds of diesel engine oxidation catalytic catalytic converter (DOC), catalytic particle catcher (CDPF) and DOC + CDPF with driving mileage.[2]The results show that with the increase of mileage, the emission reduction efficiency of DOC decreases significantly and the maintenance cost is higher. In 2021, Dai Mengbo et al. (Lin et al. 2021). conducted a study on the removal of PM10/2.5/1.0 in the sintering process of iron ore based on chemical agglomeration, and the results showed that the removal efficiency of PM10, PM2.5 and PM1.0 by chemical agglomeration method was 39.87%, 29.88% and 22.85%, respectively, with poor effect (Dai et al. 2021). In 2018, Cui Rongjian et al. (Liu et al. 2022). proposed a new particle layer emission reduction process, but it lacks the corresponding experimental basis and demonstration, and its feasibility needs to be considered (Cui et al. 2018). Patterson and Cawood first discovered the phenomenon of acoustic condensation in the laboratory, and demonstrated that sound waves can accelerate the collision and aggregation of particles, which indicates that it is possible to use sound waves to remove pollution particles suspended in the air. In fact, since the end of the twentieth century, scientists have begun to consider using the principle of sound wave condensation to remove suspended particles. After further research, it was found that the sound field coupled with the external particle field formed a joint effect of the sound field and seed nucleus, which significantly increased the collision probability of fine particles and large particles and promoted the agglomeration of sub-micron or micron particles. However, the analysis of relevant experimental data was lacking. Yan found that the solid particle acoustic wave complex field and the droplet acoustic wave complex field can increase the removal rates of fine particles in urban incinerated waste and coal by 10 and 18%, but the relationship between the relevant parameters of acoustic wave and the agglomeration effect has not been systematically studied, and the selection of the optimal operating conditions of acoustic wave lacks theoretical support. In 2002, Hou Shuangquan from Tsinghua University investigated the dissipation effect of low-frequency sound waves less than 50 Hz on water mist particles, and qualitatively concluded that lower frequency and higher sound intensity are conducive to the dissipation of water mist, which has added to the relevant theories of sound wave condensation, but there is a lack of quantitative research on the relationship between sound waves and condensation effect.

Based on the above research on acoustic agglomeration theory and particulate matter emission reduction, this paper adopted the method of combining experiment and simulation to analyze the influence of acoustic wave frequency and sound pressure level change on the agglomeration effect of particulate matter, and quantitatively simulated the process of particulate matter emission reduction through the simulation model established by dynamic equation. Compared with previous research, our research has obtained a more accurate optimal frequency interval, and a more clear explanation of the relationship between sound pressure level and the number of particles. Finally, based on the theoretical research of our team, we designed the relevant device for experimental demonstration, and the obtained results are highly consistent with the simulation conclusions, which indicates that the acoustic frequency and sound pressure level can be changed under different actual needs to achieve the best removal efficiency, and the acoustic agglomeration has a good prospect in the field of emission reduction of pollution particles.

6.2 Theoretical Study of Acoustic Wave Condensation

6.2.1 Theoretical Basis: The Mechanism of Simultaneous Motion

Sound wave condensation uses the strong sound field to cause the relative movement of polluted particles. According to the acoustic reunion related theory, the propagation of acoustic fluid media will be under the action of the sound wave regular periodic change, different scales of the particle response to the change is very different, led

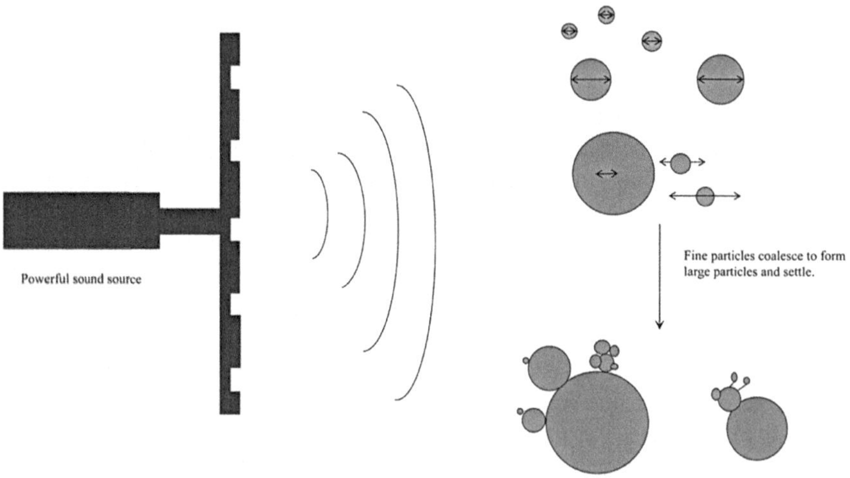

Fig. 6.1 Working principle of sound wave condensation

to the sound wave for entrainment coefficient of different pollution particles, so to strengthen the pollution particles, the movement of different particles will produce big difference. Therefore, the probability of collision and polymerization between contaminated particles is greatly increased. Under the action of multiple collision and polymerization, fine contaminated particles eventually condense into larger scale particles, which can settle under the action of gravity and no longer remain in the air (Bai et al. 2020). The working mechanism of acoustic coalescence is shown in Fig. 6.1.

6.2.2 Theoretical Calculation

In this study, the simultaneous agglomeration mechanism of sound waves was considered to eliminate contaminated particles. When the Reynolds number Re of the small particles is << 1, the small particle inertial force is small compared with the viscosity force, and its macroscopic velocity and vibration velocity are also very small. From Stokes formula can get the equation of motion of fine particles in the sound field (Hoffmann et al. 1993).

$$m_p \frac{du_p}{dt} = 6\pi \mu A_p (u_g - u_p) \tag{6.1}$$

$$\frac{du_p}{dt} = \frac{u_g - u_p}{\tau_p} \tag{6.2}$$

$$\tau_p = \frac{\rho_p d_p^2}{18\mu} \tag{6.3}$$

m_p represents the particle mass, A_p represents the particle cross-sectional area, μ represents the viscosity of the fluid, u_g, u_p refers to the particle velocity of the particle and the gas, τ_p represents the relaxation time of the particle.

From Eqs. (6.1), (6.2) we know that the relaxation time of the particle represents the speed of the particle to respond when the external force changes. The smaller the value of the relaxation time is, the smaller the inertia of the particle is, so the faster the response to the external changes is. From Eq. (6.3), the smaller the radius of the particle, the smaller the relaxation time, when the speed of the air medium changes, the small particles can respond quickly, with the medium motion faster, the large particles respond slower, with the medium motion is slower, eventually led to different particle size particles in the aluminum tube collision reunion (Fu et al. 2024).

6.3 Acoustic Wave Condensation Simulation Verification

Based on the above theoretical formula, this project adopts the control variable method to simulate the change of pollution particle concentration in the particle size range of 0–50 μm under the action of different acoustic waves, and uses the zoning method to determine the particle concentration of each different particle size. In order to obtain the precise influence of sound pressure level and acoustic frequency on the agglomeration effect of polluted particles, the team adopted self-programming programming to simulate the agglomeration effect of acoustic waves on particles under different working conditions. The action time is set to 300s, the sound pressure level is 140 dB, and the simulation frequency is 300 Hz, 500 Hz, 700 Hz, 900 Hz, 1200 Hz and 1500 Hz respectively to obtain the coagulation effect of sound waves on pollution particles, so as to obtain the optimal frequency and further obtain the optimal working condition on this basis. At the same time, this simulation also focuses on the influence of sound pressure level on the condensation effect. Under the condition that the frequency of sound wave is 800 Hz and the action time is 300s, the agglomeration effect of particles under the action of sound pressure level of 130 dB, 140 dB, 150 dB and 160 dB is respectively investigated, and the relationship between the condensation effect and sound pressure level is obtained.

6.3.1 The Condensation Effect of Particles Under the Action of Sound Waves at Different Frequencies at the Same Sound Pressure Level

The above two figures show the influence of sound waves at different frequencies on the condensation effect of polluted particles at the same sound pressure level for the same time. At the sound pressure level of 160 dB, changing the acoustic frequency, all the lines generally show the characteristics of bimodal distribution. The distribution lines have two maxima, one minimum, from 2 to 4 μm and 10 to 16 μm, respectively between reaching the peak, between 4 and 10 μm to reach the valley. According to the analysis, the high-frequency sound waves have a good condensation effect on fine particles below 6 μm, while the low-frequency sound waves have a good condensation effect on large particles above 6 μm. The above simulation results and acoustic agglomeration mechanism are highly consistent, according to the theory of acoustic reunion, the low frequency acoustic ailment coefficient is larger, and the pollution particles with the sound wave vibration ability are stronger, so the frequency sound wave of frequency sound wave is difficult to entrainment of large particles have better condensation effect, high frequency sound although entrainment coefficient is small, but the energy of low frequency acoustic more, makes the pollution particles per unit time collision times more, for lighter quality, scale smaller pollution particles, its condensation effect is better.

The particle distribution plots during the initial time and the variation of particle distribution at different frequencies are shown in Fig. 6.2. Analysis of the above images can obtain the following conclusions. The influence of acoustic frequency and the condensation effect of particles is not a simple linear relationship, but has a certain optimal frequency, so that the condensation effect of contaminated particles is the best. According to the analysis of the simulation results show that the optimal frequency is between 800 and 1100 Hz. In this frequency interval, the effect of contaminated particle condensation is optimal, and its number decreases significantly.

6.3.2 Theoretical Basis: The Mechanism of Simultaneous Motion

The sound pressure level represents the pressure generated by sound waves. The greater the sound pressure level is, the greater the amplitude of the gas in the sound field, and the greater the probability of collision and condensation between particles. Therefore, according to the relevant theory of sound wave vibration, the greater the sound pressure level is, the better the condensation effect will be.

In this simulation, the influence of sound pressure level on the condensation effect was studied by the control variable method. First of all, based on the simulation results of simulation 3.1, the sound wave frequency is controlled within the optimal

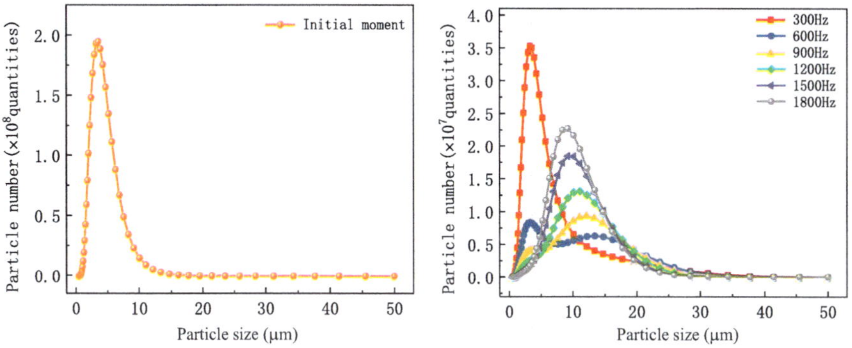

Fig. 6.2 Plot of particulate distribution at initial time (left) and change of particulate distribution at different frequencies (right)

frequency range. In this simulation, 800 Hz sound wave is selected to carry out the corresponding simulation. Change the size of the sound pressure level and explore the effect of particulate agglomeration under the conditions of 120, 140, 150 and 160 dB. The simulation results are shown in the figure below.

In this simulation, the effect of sound pressure level on the condensation effect was studied by the control variable method. Firstly, based on the results of the simulation 3.1, the acoustic wave frequency is controlled within the optimal frequency interval, and the 800 Hz sound wave is selected to carry out the corresponding simulation. Change the size of the sound pressure level to explore the effect of particle condensation under the sound pressure level of 120, 140, 150 and 160 dB. The simulation results are shown in the figure below.

The images from the simulation showed a positive correlation between the sound pressure level and the condensation efficiency of the contaminated particles. This result is highly consistent with the acoustic vibration theory, with the increase of sound pressure level, the relative velocity of different particles between the amplitude of the acoustic entrainment movement, pollution particles in an acoustic cycle of effective condensed volume also increases, the probability of collision polymerization increases accordingly, condensation effect increases.

By analyzing the simulation results, we can also conclude that although the sound pressure level is positively correlated with the condensation efficiency of the polluted particles, the increase in the condensation efficiency is not linear with the increase in the sound pressure level. For every 10 dB increase in sound pressure levels below 140 dB, the condensation efficiency achieved by polluted particles increased not significantly. When the sound pressure level is between 120 and 130 dB, the difference is small, and when the sound pressure level is greater than 140 dB, the condensation efficiency of the pollution particles increases with the sound pressure level, and the particle concentration decreases significantly. Figure 6.3 analyzes the relationship between sound pressure level and condensation time, and it can also be concluded that with the increase of sound pressure level, the improvement of condensation efficiency is more significant.

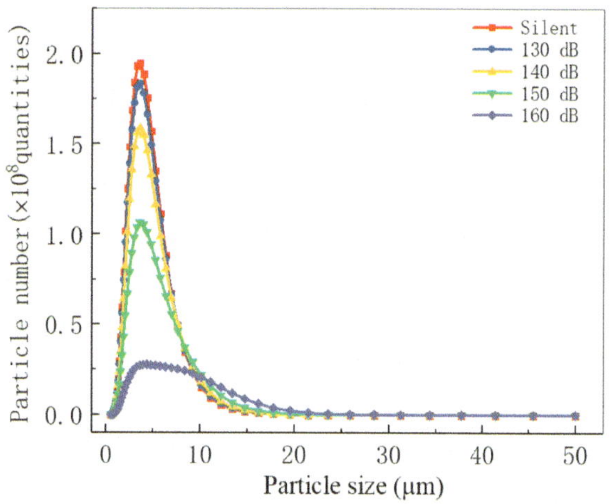

Fig. 6.3 Change curve of particle distribution under different sound pressure levels

6.4 Experiment Verification

Simulation 3.1, 3.2 presents the pollution particles condensation efficiency and sound pressure level and acoustic frequency relationship, in the process of pollution particles simulation touch, high-frequency, low-frequency acoustic waves for different scales of particles have different condensation effects, high-frequency sound wave for small scale pollution particles removal effect is good, and low-frequency acoustic wave for larger pollution particles have good condensation effect, so according to the scale of pollution particles to determine the optimal frequency and optimal condition, it also makes the acoustic condensation method has strong flexibility and maneuverability. In addition, the condensation effect of polluted particles is positively correlated with the sound pressure level, but its increase is not linear, so it is necessary to choose the appropriate sound pressure level according to the actual needs, and the sound pressure level should not be increased blindly regardless of the cost.

In order to verify the correctness of the simulation results and the feasibility of this method, we designed the acoustic wave condensation module with aluminum alloy as the main material, and conducted related experiments. In this experiment, we pumped equal mass PM2.5 particles into the acoustic condensate chamber and showed the settlement by the illumination meter value connected to the observation port.

In order to verify whether there is an optimal frequency, we controlled the sound pressure level of the sound source to keep 140 dB constant, and gradually adopted the equal interval method, with the 200, 200 Hz interval, gradually increasing the sound frequency, each group of experiments in 1, 2, 3, 4 min illumination meter, record illumination meter value stable time, when the illumination of the concentration of the settlement is complete, the shorter the time, that the faster the settlement rate, the better the condensation effect. After selecting the best frequency interval by 200Hz, the interval was further narrowed within this interval for more precise supplementary experiments to avoid the contingency of the experiment, and the results are shown in Tables 6.1 and Table 6.2.

When the frequency gradually increased from 200 to 1000 Hz, the time of the illumination value decreased from 156 to 116 s, indicating that the condensation effect is getting better and better. When the frequency increased from 1000 to 1200 Hz, the stabilization time of the illumination meter value increased by nearly 70s, and the condensation effect decreased. This trend indicates that there must be an optimal frequency, which makes the condensation effect of contaminated particles reach the best, which is highly consistent with the simulation conclusion. Through subsequent supplementary experiments, we can determine the optimal condensation frequency at around 1000 Hz, which matches the simulation results and further proves the correctness of the simulation results. Meanwhile, under the sound wave of 140 and 140 dB, which further reflects the feasibility and good prospects of this method.

Table 6.1 Initial experiment result

Initial experiment						
Frequency (Hz)	Initial illuminance/Lux	1 min	2 min	3 min	4 min	A reach stabilization time/s
200	80.3	94.9	136.9	145.7	145.7	156
400	80.0	103.7	108.0	143.6	143.6	165
600	80.9	103.7	112.3	146.1	146.1	163
800	79.5	126.3	136.5	146.5	146.5	125
1000	79.9	145.9	146.3	146.7	146.7	116
1200	79.1	85.6	118.3	130.2	146.7	186
1400	80.3	83.2	109.6	120.3	146.7	192
1600	79.6	82.1	105.6	132.6	145.9	183
1800	80.2	81.7	105.6	137.2	145.2	208

Table 6.2 Supplementary experiment result

Supplementary experiments						
Frequency (Hz)	Initial illuminance/Lux	1 min	2 min	3 min	4 min	A reach stabilization time/s
800	80.3	130.0	137.9	146.5	146.5	125
900	83.5	127.6	136.5	142.5	145.2	121
1000	81.2	129.4	140.9	146.5	146.5	116
1100	80.3	130.0	142.9	144.5	144.7	136

6.5 Summary

Simulation 3.1, 3.2 presents the pollution particles condensation efficiency and sound pressure level and acoustic frequency relationship, in the process of pollution particles simulation touch, high-frequency, low-frequency acoustic waves for different scales of particles have different condensation effects, high-frequency sound wave for small-scale pollution particles removal effect is good, and low-frequency acoustic wave for larger pollution particles have good condensation effect, so according to the scale of pollution particles to determine the optimal frequency and optimal condition, it also makes the acoustic condensation method has strong flexibility and maneuverability. In addition, the condensation effect of polluted particles is positively correlated with the sound pressure level, but its increase is not linear, so it is necessary to choose the appropriate sound pressure level according to the actual needs, and the sound pressure level should not be increased blindly regardless of the cost.

The analysis results of the simulation in this study show that the use of an acoustic agglomeration mechanism to treat PM2.5 is effective. In this process, the frequency of the sound wave and the size of the sound pressure level can be adjusted according to the actual needs to achieve the optimal removal effect. It can be seen that the treatment of contaminated particles by acoustic wave agglomeration mechanism has strong realism and operability, and has a very good application prospect and research value. Compared with the mainstream mechanism of treatment of contaminated particles, the acoustic agglomeration mechanism has the advantages of practical environmental protection and a simple process, which not only avoids the secondary pollution of chemical treatment of contaminated particles, but also uses the unutilized sound energy in many scenes as excellent catalysts to improve the energy conversion rate and utilization rate. As an emerging field, the acoustic wave agglomeration mechanism has the flexibility and advantages that other methods and mechanisms do not have, and has a good development space and application prospect.

References

Bai W, Wei J, Ni S, Shi Y (2020) Experimental study on microdroplet sedimentation under low frequency acoustic wave. J Appl Basic Eng Sci 28:247–258

Cui R, Liu H, Gao H, Gao Z (2018) Research and prospect of emission reduction of ultra-fine particulate matter in combustion flue gas. Energy and Environ Protect 40(12):123–126,131

Dai M, Ruan Z, Geng Y, Li W, Mu G, Chun T (2021) Study on removal of PM10/2.5/1.0 from chemical agglomerated iron ore sintering process. Iron and Steel Res J 33(5):385–393

Fu H, Li Y, Shen D, Ma L (2024) Process optimization of ultrasonic assisted preparation of microcapsule emulsion with enzymatic hydrolysis of crab. J Tianjin Agricul College 31(2):31–37

Hoffmann TL, Chert W, Koopmann GH, Scaroni AW, Song L (1993) Experimental and numerical analysis of bimodal acoustic agglomeration. J Vib Acoust 115(3):232–240

Liu X, Du J, Liu H, Huang Y, Wang Y, Zhang L (2022) Application progress of acoustic agglomeration coupling technology in removing fine particulate matter from flue gas. Mod Chem Ind 42(8):85–88

Lin C, Zhang G, Ma Z, Wang J, Gu H, Yuan D (2021) Study on the microscopic mechanism of acoustic agglomeration of fine particles. Acoustic Technol 40(5):587–593
Yang F, Jiang G, Liu L, Liu Y (2023) Effect of acoustic wave on combustion characteristics of pulverized coal: research progress and prospect. J China Coal Soc 48(10):3895–3911
Zhang J, Lou D, Tan P, Zhao K (2018) Effects of different reprocessing devices on emission reduction of particulate matter in diesel vehicles. J Tongji Univ (Nat Sci) 46(7):956–963

Chapter 7
System Design of Artificial Intelligence and Digital Twins for Smart Fusion Materials

Pan Liu, Jian Wang, Zhangchun Tang, and Zilong Yuan

Abstract With the continuous development of fusion technology, the demand for smart fusion materials is becoming more and more urgent. Smart fusion materials are materials with intelligent sensing and response capabilities, capable of acquiring and feeding back changes in the physical state of fusion in real time. Such materials can help scientists better understand material behaviour during fusion and provide important real-time data support. However, traditional experimental approaches often fail to meet the demand for smart fusion materials, and a new system is needed to achieve this goal. The system design proposed in this paper includes a smart material module and a digital twin module. The smart material module is used to acquire material data and interact data with the digital twin module. This module can acquire physical parameters of the material such as temperature, pressure and deformation in real time through sensors and other technologies. These data can be transmitted to the digital twin module for processing and analysis through wireless transmission or wired connection. The system achieves real-time sensing and prediction of fusion materials through the smart material module and the digital twin module. Such a system can help scientists better understand the behaviour of materials during fusion and provide important support for the further development of fusion technology.

Keyword Smart fusion materials · Artificial intelligence · Digital twin · System design

P. Liu (✉) · Z. Yuan
Tianfu Innovation Energy Establishment, Chengdu, China
e-mail: liupanccsu2020@163.com

J. Wang
Harbin Engineering University, Harbin, China

P. Liu · Z. Tang · Z. Yuan
University of Electronic Science and Technology of China, Chengdu, China

Q. Wang and A. Hosseini-Bandegharaei (eds.), *Environmental Functional Materials in Air Pollution*, Environmental Science and Engineering,
https://doi.org/10.1007/978-3-031-84571-0_7

7.1 Introduction

Fusion technology has attracted much attention as a potential energy solution (Lindl et al. 2014; Moody et al. 2014; Zylsstra and Hurricane 2022; Liu and Tang 2022; Xianjue and Cheng'an 2019). Fusion is a process that releases huge amounts of energy by fusing light elements into heavy elements, similar to the reactions that occur in the sun and stars. Unlike nuclear fission technology, the fusion process does not require the use of large quantities of radioactive materials and is therefore safer and more environmentally friendly. However, there are still many challenges to the commercialisation of fusion technology. One of them is the development and application of smart fusion materials. Fusion reactions need to be carried out under extreme conditions, such as high temperature, high pressure and strong radiation. Therefore, fusion materials need to have special properties and tolerance. Smart fusion materials are materials with self-awareness, self-adaptation and self-repair capabilities. These materials can sense the temperature, pressure, radiation and other parameters in the fusion process in real time, and automatically adjust their own performance according to this information to ensure the stability and high efficiency of the fusion reaction. In addition, smart fusion materials can also automatically repair damaged parts to extend the service life of the materials. In order to achieve the development and application of smart fusion materials, artificial intelligence and digital twin technology have been introduced into the fusion field (Degrave et al. 2022; Zylstra et al. 2021; Liu et al. 2022). Artificial intelligence can help scientists better understand the properties and behaviours of fusion materials so as to optimise material design and improve the fusion process. Digital twin technology, on the other hand, can simulate the process of fusion reactions and predict material properties and lifetimes in real time. The combination of these technologies provides strong support for the development and application of smart fusion materials. The development and application of smart fusion materials is one of them, but through the introduction of artificial intelligence and digital twin technology, it can provide important support for optimising material design and improving fusion processes.

This paper presents a system design based on artificial intelligence and digital twins to improve the efficiency and accuracy of fusion material research. The system enables real-time display and interaction of fusion materials by combining physical entities of actual fusion reactions with digital models. The digital model uses artificial intelligence algorithms to predict and optimise the performance of fusion materials. The system uses sensors and data acquisition equipment for real-time monitoring and data acquisition of physical entities in the fusion reaction. These include physical parameters such as temperature, pressure, and radiation, as well as data on the structure and properties of the material. The physical entities of the actual fusion reaction are mapped to a digital model through digital twin technology. The digital model is based on artificial intelligence algorithms to predict and optimise the performance of fusion materials. By analysing the real-time monitoring data and the prediction results of the digital model, the system can accurately judge and predict the current state of fusion physics. The artificial intelligence and digital twin system designed in this paper has achieved good results in the research of intelligent fusion materials. The system can accurately judge and predict the current state of fusion physics, and can display and interact with fusion physics entities in real time. Meanwhile, the system has good stability and reliability.

7.2 System Module Design

The system mainly consists of two parts: intelligent material module and digital twin module. The smart material module is used to obtain material data and interact with the digital twin module; the digital twin module uses deep learning algorithms to judge and save the current state of fusion physics, and uses digital twin technology to map fusion physics entities to digital twins for real-time display and interaction, and Fig. 7.1 shows the design diagram of the system modules.

The smart material module mainly includes a material library, an equation of state module and a wireless communication module. The material library is used to store the performance data of various fusion materials, including thermal conductivity, thermal expansion coefficient, melting point and so on. The equation of state module calculates the physical state of the material, such as temperature, pressure, etc., based on the material's property data. The wireless communication module is used for data interaction with the digital twin module to transmit the material property data to the digital twin module. The Smart Materials Module is a system designed based on artificial intelligence and digital twin technology to improve the efficiency

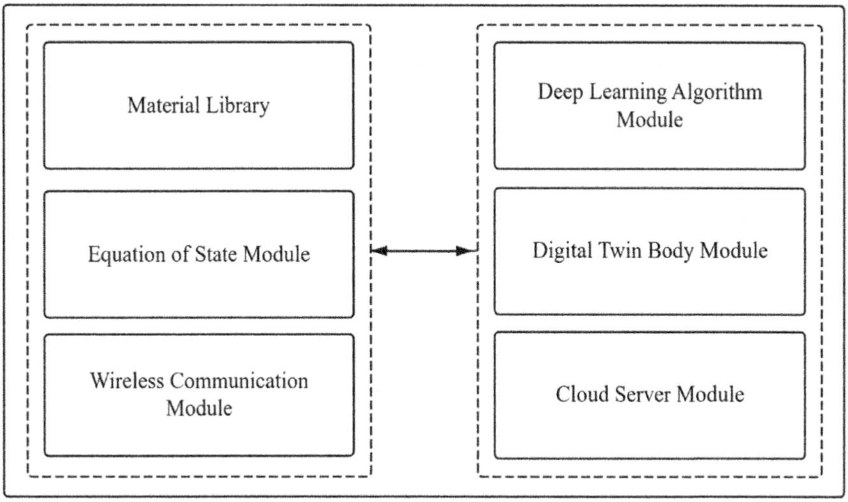

Fig. 7.1 System module design diagram

and accuracy of materials research. The module mainly consists of material library, equation of state module and wireless communication module.

The materials library is the core component of the smart materials module and is used to store the property data of various fusion materials. These property data include thermal conductivity, thermal expansion coefficient, melting point, etc., which are important indicators for assessing material properties and behaviour. The establishment of the materials library requires extensive materials testing and experimental data collection to ensure that the data in the library are accurate and reliable. Through the material library, researchers can easily query and compare the properties of different materials to provide reference for material selection and optimisation (Zylstra et al. 2021; Liu et al. 2022; Kumar et al. 2021). The equation of state module is a key component of the smart materials module, and its main function is to calculate the physical state of the material, such as temperature and pressure, based on the material's performance data. Through the equation of state module, researchers can quickly and accurately obtain information about the physical state of a material to better understand the behaviour and properties of the material. The design of the equation of state module needs to consider the properties and behavioural models of different materials to ensure the accuracy and reliability of the calculation results. The wireless communication module is the bridge between the smart material module and the digital twin module, mainly used for data interaction and transmission. Through the wireless communication module, the performance data in the material library can be easily transmitted to the digital twin module to provide data support for the establishment and optimisation of the digital twin model. At the same time, the wireless communication module can also receive commands and feedback from the digital twin module to achieve interaction and collaboration between the modules. The design of the wireless communication module needs to consider the stability and security of data transmission to ensure the accuracy and confidentiality of the data.

The digital twin module mainly includes the deep learning algorithm module, the digital twin body module and the cloud server module. The deep learning algorithm module uses deep learning algorithms to judge and predict the current state of fusion physics, including temperature, pressure and other parameters. The digital twin module uses digital twin technology to map fusion physics entities to digital twins for real-time display and interaction, which can be realised through virtual reality technology to simulate and manipulate fusion physics entities (Kumar et al. 2021; Christoph et al. 2022). The cloud server module is used to save the current state of fusion physics and provide cloud computing power to support the operation of deep learning algorithms and optimisation of models. The digital twin module plays an important role in fusion technology, which combines deep learning algorithms, digital twins and cloud servers to provide a new solution for the research and application of fusion physics.

The deep learning algorithm module is the core component of the digital twin module. By using deep learning algorithms, the module is able to accurately judge and predict the current state of fusion physics. For example, intelligent monitoring and control of the fusion reaction can be achieved by monitoring the temperature, pressure and other parameters of the fusion reaction and analysing and predicting them using deep learning algorithms. The design and optimisation of the deep learning algorithm module can improve the operational efficiency and safety of the fusion reaction. The digital twin module is a technology that maps fusion physical entities to digital twins for real-time display and interaction. Through the digital twin technology, the physical entity of the fusion reaction can be corresponded to its digital model to achieve real-time monitoring and operation of the fusion reaction. For example, the digital model of the fusion reaction can be presented to the user through virtual reality technology, so that the user can simulate and adjust the fusion reaction through interactive operation, thereby realising optimisation and improvement of the fusion reaction. The cloud server module is used to save the current state of fusion physics and provide cloud computing power to support the operation of deep learning algorithms and optimisation of the model (Schenck and Fox 2017; Liu and Yuan 2024; Xiong et al. 2023). The operation of the fusion reaction generates a large amount of data, which can be stored and managed by the cloud server for subsequent analyses and applications. At the same time, the cloud server can also provide powerful computing power to support the operation of deep learning algorithms and model optimisation, thus improving the operational efficiency and performance of the fusion reaction.

7.3 System Workflow

The workflow of this system is as follows:

(1) The smart material module transmits material property data to the digital twin module via a wireless communication module. Smart material modules typically contain sensors to collect material property data such as temperature, pressure, and strain. This data is essential for understanding the behaviour and properties of materials. However, traditional data collection methods often require manual intervention, which is not only time-consuming and labour-intensive, but also prone to errors. To solve this problem, wireless communication modules have been introduced into smart material modules to enable real-time data transmission. The wireless communication module uses wireless technologies, such as Bluetooth, Wi-Fi or LoRa, to transmit the performance data collected by the smart material module to the digital twin module. The wireless communication module is characterised by high speed transmission, low power consumption and long distance transmission, which enables a seamless connection between the smart material module and the digital twin module. The digital twin module is the core module in the fusion physics research system, which establishes the digital twin of the material by receiving the performance data from the smart material module. The digital twin is a virtual copy of the material that simulates the behaviour and properties of the material in different environments. By comparing and analysing the material with the actual material, researchers can better understand the material's response and adaptability to optimise the fusion physics process. Through the collaboration of the Smart Materials Module and the Wireless Communication Module, the material's performance data can be transmitted to the

Digital Twin Module in real time, providing researchers with timely and accurate data support. This seamless connection and real-time data transmission greatly improves the research efficiency and reduces the error of manual intervention, bringing new possibilities for fusion physics research.

(2) The Deep Learning Algorithm module uses the transmitted material data to determine and save the current state of the fusion physics. The deep learning algorithm module needs training data to learn the relationship between different material properties and fusion physics states. These training data can include information such as the structure, composition, temperature, and density of the material, as well as the corresponding fusion physics state, such as the temperature, density, and stability of the plasma. With a large number of training samples, the deep learning algorithm module can learn the complex relationships between these features and build a corresponding model (Jiang, et al. 2022; Hu, et al. 2405). Here the reset unit R and update unit U are used to control the parameter flow of the net in the model:

$$Rs_t = \delta(W_{r1}M_t + W_{r2}s_{t-1} + B_r) \tag{7.1}$$

$$Us_t = \delta(W_{u1}M_t + W_{u2}s_{t-1} + B_u) \tag{7.2}$$

where δ is a nonlinear function, St is the output value of the network at time t. W is the parameter to be optimized for the network, and B is the bias parameter of the network. M is the two-dimensional parameter matrix generated by data preprocessing.

The Deep Learning Algorithm module can use the transmitted material data to determine the current state of fusion physics. When new material data is transmitted to the deep learning algorithm module, the module automatically extracts the features in it and makes a judgement through the previously learned model. For example, the module can predict the current plasma temperature, density, and other state parameters based on the transmitted material temperature and density data. In this way, the deep learning algorithm module can monitor and save the current state of fusion physics in real time. In order to improve the accuracy and stability of the deep learning algorithm module, it is also necessary to update and optimise the module on a regular basis. As the research continues, new training data can be added gradually to cover more material properties and fusion physics states. At the same time, the model can also be adjusted and improved to enhance its predictive ability and adaptability.

(3) The digital twin module uses digital twin technology to map fusion physical entities into digital twins for real-time display and interaction. The digital twin module uses digital twin technology to convert various parameters and properties of fusion physical entities into digital form and display them in the digital twin in real time. Through the digital twin module, changes in the morphology, temperature, pressure and other parameters of the fusion physical entity can be observed and interacted with in real time. This allows researchers to understand the state of fusion physical entities more intuitively and take appropriate control measures in time (Chen et al. 2020). The application of the digital twin module also allows the prediction of the response and behaviour of fusion physical entities by simulating

different operational and disturbance scenarios. In this paper the mapping variable X is an M × N matrix.

$$X = \begin{bmatrix} x_{11} & x_{12} & \cdots & x_{1N} \\ x_{21} & x_{22} & \cdots & x_{2N} \\ \vdots & \vdots & \ddots & \vdots \\ x_{M1} & x_{M2} & \cdots & x_{MN} \end{bmatrix} \tag{7.3}$$

By controlling and manipulating the digital twin, researchers can simulate different experimental conditions and control strategies in order to better optimise the performance and stability of fusion physics entities. The digital twin module can also be linked with other modules to achieve more functionality. For example, combining the digital twin module with the control module enables real-time control and regulation of fusion physical entities. Combining the digital twin module with the optimisation module can achieve the optimisation of the performance and stability of fusion physical entities.

The cloud server module saves the current state of fusion physics and provides cloud computing power to support the operation of deep learning algorithms and optimisation of models. The cloud server module records and saves the state and behaviour of fusion physics entities in real time. By linking with other sensors and devices, it is able to acquire a large amount of real-time data and store it in the cloud to provide a data base for subsequent analyses and simulation calculations. Deep learning algorithms play an important role in fusion physics research, and can extract useful features and patterns from large amounts of data (Hafner, et al. 2301). The cloud server module provides powerful computing power that can support the operation of deep learning algorithms and accelerate the process of data processing and pattern recognition. The cloud server module can also optimise the deep learning models through the power of cloud computing. Through parallel computing and distributed computing, the training and tuning process of the model can be accelerated to improve the accuracy and efficiency of the model.

7.4 Physical Design and Machine Experiment Discussion

Fusion technology is an energy solution with great potential, but its complexity and high risk make accurate prediction and real-time monitoring of the physical behaviour of fusion an important and difficult task. To address this challenge, a comprehensive control and optimisation system has been developed that integrates a smart materials module and a digital twin module to enable the comprehensive management of fusion physics through the selection of the most suitable materials from the materials library, the prediction of material properties and behaviour, real-time monitoring and adaptive control.

Table 7.1 The comparison of physical parameters

Physical parameters	Numerical value	Range of error (%)
Velocity	2.7×10^7 cm/s	5
Density	10 g/cc	1
Temperature	3.5 keV	3
N eutron yield	1×10^{18}	8

The system is first used to test the state of known fusion physics and to verify the system's prediction results. The experimental results show that the system is able to accurately judge and predict the current state of fusion physics with a prediction error within acceptable limits. This lays the foundation for the reliability of subsequent experiments. Table 7.1 shows the comparison of physical parameters.

Next, a series of real-time monitoring experiments were conducted to display and interact with fusion physics entities in real time through the system in order to explore its feasibility in practical applications. The results of the experiments show that the system is able to monitor the changes in fusion physics in real time and can provide accurate data and image displays. The system also allows users to perform interactive operations to achieve real-time regulation and optimisation of fusion physics by adjusting parameters and control strategies. These results demonstrate the reliability and flexibility of the system in practical applications. Table 7.2 shows the real-time comparison of the data.

The system was also tested for stability. The system was found to have good stability and reliability through long-time operation and experiments under different working conditions. The system was able to work properly both in regular operation and under extreme conditions, and was able to provide accurate prediction and monitoring results consistently. In summary, a series of experimental results have been produced by testing and validating the system. These results show that the system is able to accurately judge and predict the current state of fusion physics, and to display and interact with fusion physics entities in real time. Meanwhile, the system has good stability and reliability. These results provide strong support for the development of fusion technology and are expected to play an important role in the field of intelligent fusion materials in the future.

Table 7.2 The real-time comparison of the data

Physical position (cm)	Physical parameters	Numerical value
1	Density	10 g/cc
1	Temperature	3.5 keV
2	Density	5 g/cc
2	Temperature	1 keV
5	Density	0 g/cc
5	Temperature	0.01 keV

7.5 Conclusion

In this paper, a system design of artificial intelligence and digital twin for smart fusion materials is proposed, implemented and verified. The experimental results show that it is consistent with the experimental results in terms of physical parameters, and there is a significant improvement in terms of neutron yield to a certain ablation depth. Meanwhile, in terms of images are transmitted in real time, and the system has high accuracy, stability and reliability when tested in long time running and non-stop condition. However, due to the limitation of time and resources, the research in this paper still has some shortcomings. Future work can further improve the function and performance of the system and conduct more in-depth research. These results provide strong support for the development of fusion technology, and are an important step towards the realisation of controlled and reliable fusion reactions lease note that the first paragraph of a section or subsection is not indented. The first paragraphs that follows a table, figure, equation etc. does not have an indent, either.

Acknowledgements This work was supported by the Natural Science Foundation of Sichuan under 2024NSFSC0503.

References

Chen D, Zhou B, Koltun V et al (2020) Learning by cheating. In: Conference on robot learning. PMLR, pp 66–75

Christoph M, Andersen J, Spasovski J et al (2022) Understanding decentralization of decision-making power in proof-of stake blockchains: an agent-based simulation approach. European Journal of Information Systems (EJIS), pp 1–20

Degrave J, Felici F, Buchli J et al (2022) Magnetic control of takamak plasmas through deep reinforcement learning. Nature 602:414–419

Hafner D et al (2023) Mastering diverse domains through world models. arXiv:2301.04104

Edward SH et al (2024) Privileged sensing scaffolds reinforcement learning. arXiv:2405.14853v1

Jiang M et al (2022) Artificial intelligence and digital twin system and method for smart clothing. Chinese invention patent. CN114678097A

Kumar A et al (2021) RMA: rapid motor adaptation for legged robots. Robot: Sci Syst (RSS). https://doi.org/10.15607/RSS.2021.XVII.011

Lindl J, Landen O, Edwards J et al (2014) Review of the national ignition campaign 2009–2012. Phys Plasmas 21(2):020501

Liu P, Zhou YZ, Xiong WB et al (2022) Big data-aided study of the physical process. J Phys: Conf Ser 2381:012064

Liu P, Yuan Z et al (2024) Metaverse drive Z-FFR physical design. Proc SPIE 13071:1307139

Liu P, Tang Z (2022) Physical design of local-volume ignition for inertial confinement fusion. IEEE Computer Society. CPS. ITCEM, pp 10–15

Moody JD, Callahan DA, Hinkel DE et al (2014) Progress in hohlraum physics for the national ignition facility. Phys Plasmas 21(5):056317

Schenck C, Fox D (2017) Visual closed-loop control for pouring liquids. In: 2017 IEEE international conference on robotics and automation (ICRA). IEEE, pp 2629–2636

Xianjue P, Cheng'an L (2019) Future nuclear energy and Z-pinch driven fusion fission hybrid reactor. Defense Industry Press, Beijing

Xiong W, Tang Z, Liu P et al (2023) Artificial intelligence-assisted design of load-driver end docking in Z-FFR fusion explosion chambers. Proc SPIE 12801:128015A

Zylsstra AB, Hurricane OA et al (2022) Burning plasma achieved in inertial fusion. Nature 601:542–548

Zylstra AB et al (2021) Record energetics for an inertial fusion implosion at NIF. Phys Rev Lett 126:025001

Chapter 8
Modeling Analysis of the Adsorption–Desorption Process of Activated Carbon Canister in On-board Refueling Vapor Recovery System

Daming Liu, Yaqi Li, and Xudong Zhen

Abstract This study investigates the activated carbon canister within the On-board Refueling Vapor Recovery (ORVR) system for gasoline vehicles. A one-dimensional hydrocarbon adsorption simulation model, based on the Dubinin-Astakhov and linear driving force models, was validated against experimental data. The model demonstrates adaptability across various carbon types and canister dimensions. At the 2 g breakthrough point, butane distribution is uneven, particularly near the outlet, leaving some activated carbon underutilized. High desorption rates reduce residual butane, but slower rates prove more effective with equivalent inlet air volumes. Significant temperature fluctuations during adsorption/desorption impact reaction kinetics. Two strategies are proposed to enhance canister performance: (1) multi-port designs for uniform butane-activated carbon contact; (2) increased desorption rates with sufficient air supply. This research provides a theoretical basis and practical guidance for ORVR activated carbon canister design and optimization.

Keywords ORVR system · Activated carbon canister · Adsorption model · Desorption efficiency · Temperature control

---

D. Liu (✉) · X. Zhen
School of Automobile and Transportation, Tianjin University of Technology and Education, Tianjin, China
e-mail: liudaming2001@163.com

Y. Li
Department of Railway Engineering, Tianjin Railway Technical and Vocational College, Tianjin, China

Q. Wang and A. Hosseini-Bandegharaei (eds.), *Environmental Functional Materials in Air Pollution*, Environmental Science and Engineering,
https://doi.org/10.1007/978-3-031-84571-0_8

8.1 Introduction

Refueling emissions from gasoline vehicles contribute significantly to pollution, prompting the adoption of On-board Refueling Vapor Recovery Systems (ORVR) to meet stringent emission standards like China VI and Euro VII (Ministry of Ecology and Environment … 2016; European Parliament News 2024). However, the rise of plug-in hybrid and electric vehicles challenges ORVR efficiency due to reduced canister desorption (Wu and He 2024). The activated carbon canister (AC canister) is crucial in ORVR, influencing system efficiency and emission control.

Sircar and Hufton (2000) analyzed the Linear Driving Force (LDF) model for gas adsorption kinetics, laying a theoretical foundation for AC canister design. Lavoie et al. (1996) enhanced simulation accuracy by developing and refining a one-dimensional canister adsorption model (CANMOD), incorporating mass transfer and thermal effects. Smith et al. (2015) integrated an engine system model with an EVAP sub-model, detailing AC canister kinetics for ORVR optimization. Östermark (2018) used CFD to study AC canister's non-isothermal behavior and heating's impact on desorption. Romagnuolo et al. (2022, 2020) analyzed AC canister behavior during loading and purging, developing a non-isothermal model considering temperature effects. Ling et al. (2022) created a real-time desorption flow rate prediction model for EVAP system calibration. Chen et al. (2024a, 2024b) characterized the adsorption and deterioration in carbon canisters under refueling, while also examining temperature's effect on adsorption efficiency.

While previous studies have made significant progress, certain limitations persist under the new Euro VII emissions standards. This study focuses on the AC canister in ORVR, developing a one-dimensional HC adsorption model based on the Dubinin-Astakhov and LDF models, validated against experiments, to support ORVR design and optimization.

8.2 Mathematical Models

8.2.1 Adsorption Rate and Linear Driving Force Model

The Linear Driving Force (LDF) model is utilized to describe the rate at which an adsorbate, whether a pure gas or a mixture containing inert components, adsorbs onto the surface of an adsorbent particle (Lavoie et al. 1996). Due to its simplicity, analytical nature, and clear physical significance, this model is frequently applied in the context of adsorption column modeling. According to the LDF model, the instantaneous rate of adsorption of a single adsorbate into the adsorbent particle is given by the following formula:

$$\frac{d\overline{C}(t)}{dt} = k_L[\overline{C^*}(t) - \overline{C}(t)] \tag{8.1}$$

where, $\overline{C}(t)$ denotes the average concentration of the adsorbate within the adsorbent particle at time t (moles per unit volume), and $\overline{C^*}(t)$ denotes the adsorbate concentration at equilibrium. Assuming uniform temperature throughout the adsorbent particle, the variable k_L represents the effective LDF mass transfer coefficient at adsorbate load $\overline{C}$ and temperature T. By expressing the adsorbed state concentration of the adsorbate as a dimensionless coverage, the LDF model formulation is given by:

$$\frac{d\overline{\theta}(t)}{dt} = k_L[\overline{\theta^*}(t) - \overline{\theta}(t)] \tag{8.2}$$

8.2.2 *Isotherm Adsorption Lines and the Dubinin-Astakhov Adsorption Model*

By using the isothermal adsorption lines of activated carbon obtained from experiments and curve fitting, the parameters of the Dubinin-Astakhov (D-A) model based on micropore filling theory can be obtained, thereby characterizing the adsorption properties of activated carbon.

$$\theta = \frac{V}{V*} = A^* e^{-\left(\frac{f}{f_1}\right)^n}, f \geq f^* \tag{8.3}$$

where, θ, V, $V*$, f represent the coverage, the adsorption volume, the equilibrium adsorption volume, the adsorption potential per unit volume of the liquid adsorbate, respectively. A^*, f_1, n are model parameters, and f^* is the boundary value of the fitting curve.

According to Polanyi's adsorption potential theory, if the adsorption temperature is significantly lower than the critical temperature of the gas, and assuming the gas behaves as an ideal gas and the adsorbed phase is an incompressible saturated liquid, then the adsorption potential per unit volume of the liquid adsorbate can be expressed as

$$f \equiv \frac{\rho_L RT}{W} \ln\left(\frac{p_{sat}}{p_{vap}}\right) \tag{8.4}$$

where, ρ_L, T, W, R, p_{vap}, p_{sat} denote the liquid density of the adsorbate, the carbon bed temperature,the molecular weight of the adsorbate,the ideal gas constant, the vapor pressure of the adsorbate in the gas phase, the saturated vapor pressure of the adsorbate, respectively. p_{sat} can be derived using the Antoine equation.

At the state of adsorption equilibrium, the adsorption potential is utilized to correlate the normalized coverage θ with the partial pressure p_{vap} of the gaseous adsorbate butane C_4H_{10}. Finally, the following expression can be derived:

$$\frac{d\overline{C}(t)}{dt} = k_L[\overline{C^*}(t) - \overline{C}(t)] = k_L[k_{\mathrm{DES}} \cdot \frac{p_{sat}}{RT} e^{-\left(\frac{f_1 W}{\rho_L RT}\right)\cdot\left(-\ln\left(\frac{\theta(t)}{A^*}\right)\right)^{\frac{1}{n}}} - \overline{C}(t)] \quad (8.5)$$

It should be noted that to facilitate model calibration, a calibration parameter k_{DES} has been incorporated into the above equation to regulate the adsorbed quantity in the actual equilibrium state.

Parameters of interest that require experimental determination encompass the effective mass transfer coefficient k_L of the linear driving force model and the adsorption coefficient k_{DES} at equilibrium. Furthermore, to account for the temperature variations occurring during adsorption, the enthalpy of adsorption E_f must also be specified.

8.3 Simulation Methods

This study utilizes the one-dimensional numerical simulation software GT-suite to address the continuity, momentum, energy, and component equations within fluid dynamics problems. The porous medium carbon bed, formed by activated carbon, is modeled using the CatalystBrick component. Subsequently, the equations that delineate the adsorption and desorption processes of activated carbon are integrated into the computational model via the SurfaceReactions component. The parameters associated with the activated carbon are based on the experimental data from Lavoie et al. (1996) and Östermark (2018), as detailed in Table 8.1.

In compliance with the evaporative emission testing standards outlined in China VI vehicle emission standards, the adsorption process within the AC canister utilizes a mixture of 50% butane (C4H10) and 50% nitrogen (N2) by volume. Butane is fed into the adsorption port at a mass flow rate of 40 g/h, continuing until the cumulative butane gas emission at the air outlet of the canister reaches the threshold of 2 g, which is identified as the 2 g breakthrough point. During desorption, compressed air is introduced at a rate of 25 L/min. Various flow rates will be simulated to assess their impact on the adsorption and desorption processes.

To model the AC canister in one dimension, the original canister is equated to an adsorption column of identical volume, based on its cross-sectional area. The effects of different canister volumes on flow and heat transfer are depicted through the effective mass transfer coefficient k_L, the equilibrium adsorption coefficient k_{DES}, and the adsorption enthalpy E_f, as detailed in Table 8.2.

Table 8.1 Parameters of the D-A model for BAX950 and BAX1500 activated carbons

	BAX950	BAX1500
A^*	1.076	1
f_1	85	69.5
N	1.2	1.43

Table 8.2 Calibration parameters for various AC canister models

	k_L	k_{DES}	E_f
Östermark (2018)	0.0025	0.5	−155
This study	1.76E−04	0.3046	−155

Table 8.3 Experimental and simulated operating conditions

	Östermark (2018)		This study	
	Adsorption	Desorption	Adsorption	Desorption
AC canister volume	1L(3846.5 mm^2 × 264 mm)		2L(3846.5 mm^2 × 528 mm)	
Activated carbon Type	BAX1500		BAX1500	
Inlet working fluid	50% Butane 50% Nitrogen (by volume)	Air	50% Butane 50% Nitrogen (by volume)	Air
Inlet flow rate	40 g/h (Butane)	23 L/min	40 g/h (Butane)	15 L/min
Inlet temperature (K)	296	296	300	300

The model's universality was initially verified by comparing its predictions with experimental results from Östermark (2018). Further validation was achieved through a 2g test and desorption test on a 2-L AC canister. The experimental conditions corresponding to Östermark (2018) and this study are presented in Table 8.3. Figure 8.1 illustrates the one-dimensional adsorption/desorption model of the AC canister that was constructed.

8.4 Results and Discussion

8.4.1 Model Validation

The model validation against experimental results, as shown in Fig. 8.2, indicates that the one-dimensional simulation model's robust adaptability across various activated carbon types and canister volumes. Under adsorption conditions, the simulation aligns closely with experimental findings, showing an initial constant adsorption rate with the adsorption amount linearly increasing over time. As the process advances, a notable deceleration in the rate is observed due to the saturation of adsorption sites, leading to a plateau in the adsorption amount, indicative of the AC canister reaching the 2 g breakthrough point, as illustrated in Fig. 8.2a. In contrast, Fig. 8.2c indicates an abrupt cessation of the adsorption experiment upon reaching this threshold, hence the absence of a distinct plateau in the curve. For desorption conditions, while the

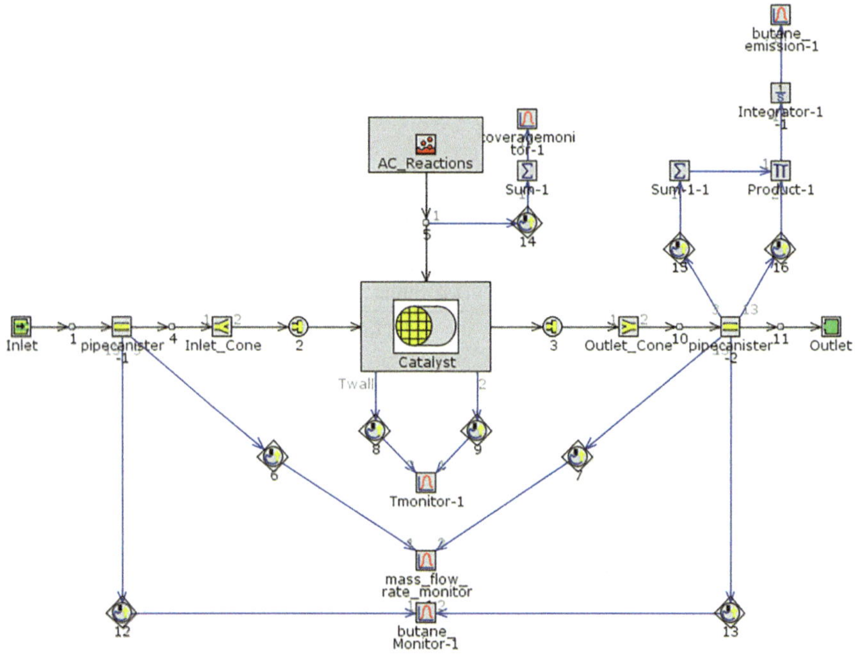

Fig. 8.1. 1-D adsorption/desorption model of the AC canister

simulation's general trend mirrors experimental results, it overshoots the desorption rate in the initial and middle phases but underestimates it in the later stages, suggesting an overestimation of the activated carbon's desorption capacity at saturation and an underestimation in the low coverage phase. Further research is warranted to dissect these discrepancies and refine the model's accuracy in simulating the desorption process.

8.4.2 Results Analysis

Utilizing the simulation model developed in this paper, an in-depth analysis was conducted to explore the residual butane mass and spatial distribution within the AC canister at various desorption rates, and the spatial and temporal variation of gas temperature during the adsorption and desorption processes. During the adsorption phase, the test was conducted in accordance with the 2 g breakthrough test conditions as defined by China VI vehicle emission standards. Once the emission from the canister reached 2 g at about 6.5 h (23,400 s), the desorption test commenced. The test evaluated the effects of three desorption rates -15 L/min, 20 L/min, and 25 L/min - on the desorption process within the canister.

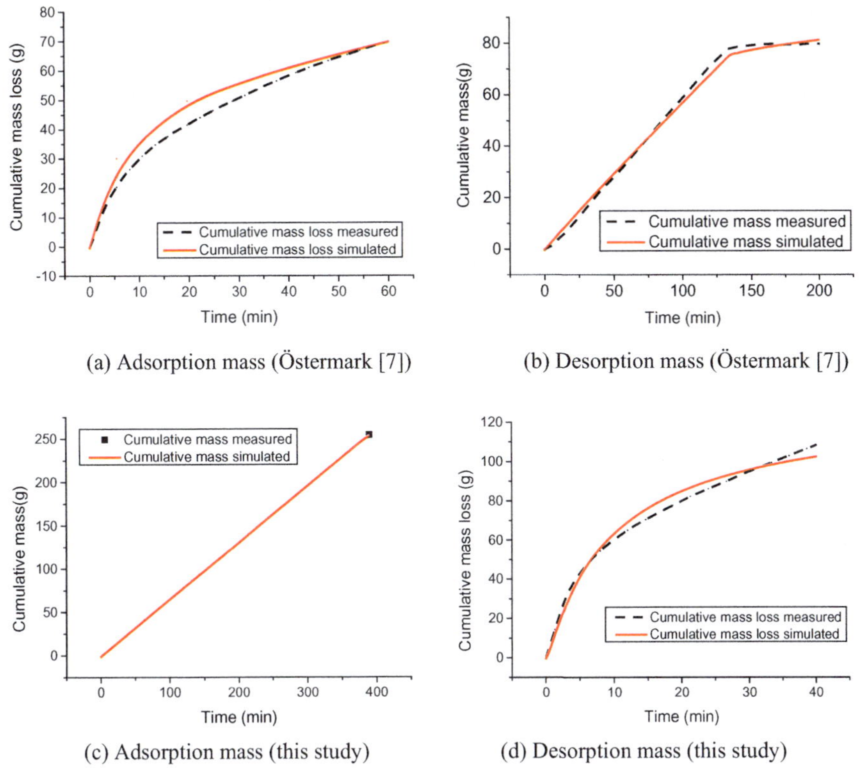

Fig. 8.2 Experimental validation of the 1-D simulation model of the AC canister

Figure 8.3 depicts the temporal variation in the total adsorbed amount of butane within the AC canister during the 2 g breakthrough test and desorption test. At the 2 g breakthrough point in the adsorption phase, the canister's maximum adsorption capacity for butane is determined to be 253.02 g. Upon transitioning to the desorption phase, it is evident that at identical desorption times, a higher desorption rate results in a lower residual butane content within the canister. It is, however, noteworthy that when an equivalent volume of compressed air is introduced—such as 300 BV, which equates to 300 times bed volume of the canister—the lower desorption rate actually corresponds to a reduced residual butane quantity, as illustrated in Fig. 8.4.

Figure 8.5 illustrates the distribution of residual butane (in terms of adsorption site coverage) along the axial direction within the AC canister, which is divided into 132 grid sections from the inlet (grid 1) to the outlet (grid 132). During the adsorption phase, the grid volumes near the inlet rapidly achieve a 90% coverage, leading to a subsequent deceleration in the adsorption rate. As butane in the inlet gas is progressively adsorbed and its concentration diminishes, the central and outlet-adjacent grid volumes also quickly reach a 57% coverage, after which the adsorption rate markedly decreases. At the critical 2 g breakthrough point, only the first 30 grid

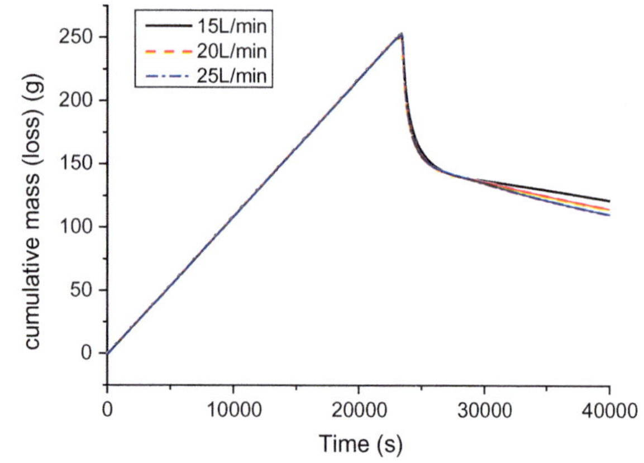

Fig. 8.3 Total butane adsorbed in the AC canister

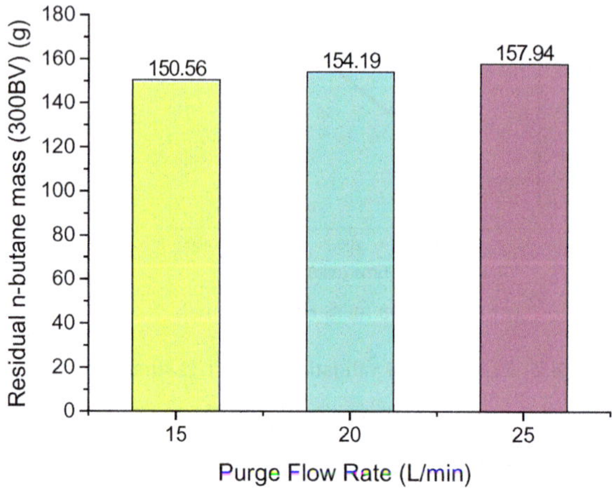

Fig. 8.4 Residual butane in the canister after 300 BV of purge airflow

volumes near the inlet exceed a 75% coverage rate, with only the first 18 reaching over 90%. This suggests a highly uneven distribution of adsorbed butane, with a significant portion of the activated carbon near the outlet end being underutilized.

In the desorption phase, at a given desorption time, a higher desorption rate results in a lower residual butane concentration within the same grid volume. For instance, consider the desorption time of 23,400 s, corresponding to the point when 300 BV of compressed air is introduced at a flow rate of 15 L/min. If the desorption rate is increased from 15 L/min to 25 L/min, the residual butane coverage in grids 1, 30, and 132 decreases from 0.447, 0.548, and 0.319 to 0.444, 0.513, and 0.291, respectively. Conversely, if the total amount of desorption air remains constant (e.g., 300 BV), a lower desorption rate actually leads to a lower coverage, indicating higher desorption efficiency. This effect is particularly noticeable near the inlet and outlet regions of the canister, as illustrated in Fig. 8..6.

Fig. 8.5 Distribution of residual butane (normalized coverage) in AC canister

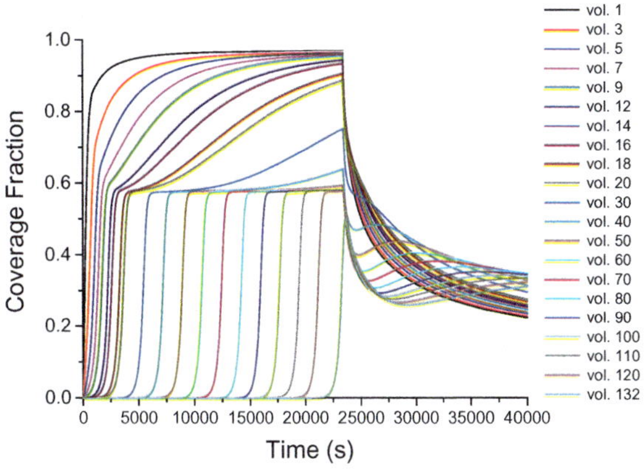

(a) Purge flow rate 15 L/min

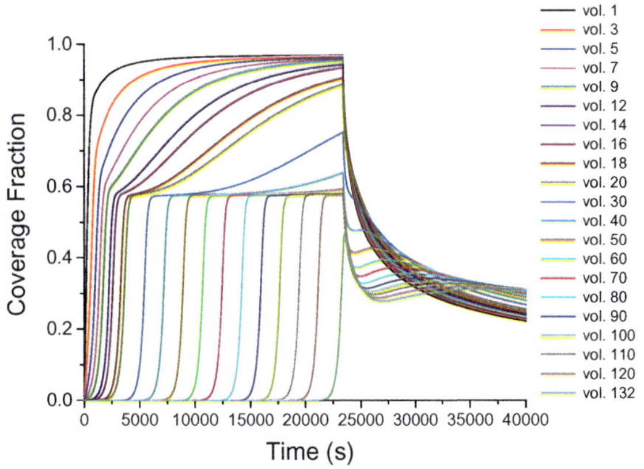

(b) Purge flow rate 25 L/min

Fig. 8.6 Distribution of butane in AC canister after 300 BV of purge airflow

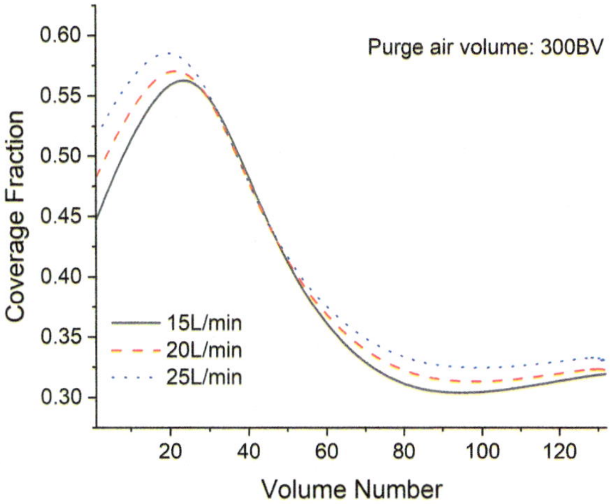

8.5 Conclusions

This study developed a one-dimensional simulation model for hydrocarbon (HC) adsorption in the AC canister of the On-board Refueling Vapor Recovery (ORVR) system, utilizing the Dubinin-Astakhov adsorption model and the linear driving force model. The model was validated against experimental data, demonstrating strong adaptability across various types of activated carbon and different canister volumes. The analysis of the adsorption and desorption processes within the canister revealed that, at the 2 g breakthrough point, the distribution of adsorbed butane remains uneven, with a significant portion of the activated carbon near the outlet still unused. At the same desorption time, higher desorption rates result in less residual butane within the canister. However, when an equivalent volume of air is introduced (such as 300 BV), lower desorption rates actually result in less residual butane. Additionally, the adsorption and desorption processes are associated with changes in adsorption enthalpy, leading to significant temperature variations during these processes.

Based on these findings, the performance of the AC canister—specifically its butane adsorption capacity—can be enhanced in two ways: (1) Implementing a multi-inlet design would allow the inlet butane gas to more evenly contact the activated carbon, thereby improving carbon utilization and increasing adsorption efficiency. (2) During the desorption phase, increasing the desorption rate as much as possible, while maintaining the air volume, can enhance desorption efficiency.

Acknowledgements The study is financially supported by the National Natural Science Fund of China (No. 51706155) and The School-Level Scientific Research Project of Tianjin University of Technology and Education (No. K15-20).

References

Ministry of Ecology and Environment of the People's Republic of China: GB18352.6-2016 Limits and measurement methods for emissions from light-duty vehicles (CHINA 6). Standards Press of China, Beijing (2016)

European Parliament News, https://www.europarl.europa.eu/news/en/press-room/20240308IPR19017/euro-7-parliament-adopts-measures-to-reduce-road-transport-emissions. Last accessed 13 March 2024

Wu X, He R (2024) Modeling and influence factors analysis of refueling emissions for plug-in hybrid electric vehicles. J Energy Res Technol 146(2):021702

Sircar S, Hufton JR (2000) Why does the linear driving force model for adsorption kinetics work? Adsorption 6(2):137–147

Lavoie GA, Johnson PJ, Hood JF (1996) Carbon canister modeling for evaporative emissions: adsorption and thermal effects. SAE Trans 105:1028–1049

Smith L, Hussain A, Pautasso E et al (2015) EVAP system fluid-dynamics and chemistry modelling for ems purge control development and optimization. In: International congress: SIA powertrain—versailles 2015. SIA, Paris, France, pp 1–10

Östermark E (2018) Modelling of a purge heated carbon canister using CFD. Gothenburg, Sweden; Chalmers University of Technology

Romagnuolo L, Frosina E, Fortunato F et al (2022) 1D model for n-butane adsorption and thermal variation for EVAP canister of gasoline-fueled vehicles: validation with experimental results and DFSS optimization. Appl Therm Eng 209:118267

Romagnuolo L, Ponticelli M, Fortunato F, et al (2020) Experimental adsorption and desorption characterization of a gasoline-fueled vehicle carbon canister for European application filled with n-butane and nitrogen mixtures. In: 75th national ATI congress. EDP Sciences, Rome, Italy, pp 06016

Ling J, Chen Q, Liu D, et al (2022) Real-time desorption flow prediction of automobile carbon canister for virtual calibration. In: 2022 IEEE 2nd international conference on electronic technology, communication and information (ICETCI). IEEE, Changchun, China, pp 56–60

Chen Z, Chen Q, Zhang X et al (2024a) Characterization of activated carbon adsorption and deterioration in carbon canisters during vehicle refueling. Fuel 374:132464

Chen Z, Chen Q, He R (2024b) Effect of ambient temperature on the adsorption performance of automotive activated carbon: an experimental and molecular simulation study. Powder Technol 443:119851

Chapter 9
Novel Insights into Anaerobic Digestion of Different Compositions in Lignocellulosic Biomass: Focus on Biomethane Production Potential, Kinetic Analysis, and Microbial Community Characteristics

Xiteng Chen, Xinzi Wang, and Ziqing Li

Abstract In order to research biomethane production potential, and analyze microbial community characteristics of lignocellulosic biomass, microcrystalline cellulose, xylan, alkaline lignin, and corn straw were mixed with anaerobic granular sludge for anaerobic digestion. The cumulative methane production of MC, XY, and CS system was 285.23 ± 5.17 mL/ VS, 370.83 ± 14.60 mL/g VS, 172.06 ± 8.56 mL/g VS respectively. Cellulose and hemicellulose can be converted by microorganisms into biomethane. The biodegradability and dynamic model fitting results also revealed the anaerobic lag time of MC system was longer than XY, and CS system. In addition, the fermentation cycle of MC also was longer than XY and CS. *T78*, *vadinCA02*, *Clostridium* and *Bacteroides* played important roles in hydrolytic-acidification, and especially the relative abundance of *Clostridium*, a genus of cellulose-degrading bacteria, increased to 5.65% (MC), 3.65% (XY) and 2.84% (CS) respectively. Moreover, *Methanosaeta* and *Methanobacterium* were the two main anaerobic archaea in MC, XY and CS anaerobic digestion systems.

Keywords Lignocellulosic biomass · Anaerobic digestion · Biomethane production potential · Microbial community characteristics

X. Chen (✉) · X. Wang
Beijing University of Chemical Technology, Beijing, China
e-mail: hjghcxt415@126.com

Z. Li
Chemical Industry Press Co., Ltd., Beijing, China

Q. Wang and A. Hosseini-Bandegharaei (eds.), *Environmental Functional Materials in Air Pollution*, Environmental Science and Engineering,
https://doi.org/10.1007/978-3-031-84571-0_9

9.1 Introduction

Anaerobic digestion (AD) has been widely recognized as an important technology for green recycling, especially for the treatment of lignocellulosic biomass waste. Lignocellulosic biomass primarily consists of lignin, cellulose, and hemicellulose. But primarily due to the recalcitrant nature of biomass originating from the complex structure of the lignocellulose, only 20–30% of the biomass (primarily extractives and hemicellulose) was converted into biomethane during AD process (Isikgor and Becer 2015; Kucharska et al. 2018). Thus, there is a critical need to enhance the digestibility of such lignocellulosic feedstock to improve the overall methane yield. Pretreatment methods such as physical, chemical, physicochemical, or biological pretreatment methods enhancing the digestibility of lignocellulosic feedstock and improving the overall biomethane production had been studied (Wang et al. 2020; Konan et al. 2024). Du et al. studied the physical–chemical combined pretreatment for rice straw, and achieved 24% more biogas production compared to that of the control (Du et al. 2019). In addition to the nature of the material itself, anaerobic digestion system also depends on microbial communities to convert lignocellulosic biomass into biomethane. The ability of microorganisms to utilize different components of lignocellulose had been investigated. Li et al. simulated cellulose using microcrystalline cellulose and α-cellulose, represented hemicellulose with xylan, galactomannan, and arabinogalactan, and used alkali lignin to simulate lignin (Li et al. 2018a). The study indicated that co-fermentation of lignin with cellulose and hemicellulose inhibits biogas production, highlighting that the primary sources of biogas are cellulose and hemicellulose. While the hydrolysis rate of hemicellulose is faster than that of cellulose, the methane production potential of cellulose remains higher.

Many studies had conducted anaerobic fermentation of different components of lignocellulose to produce biogas, however not defined the biomethane production potential. And simultaneously, the current trend in treating lignocellulosic biomass still focus on pretreatment technology to break apart the complex structure of lignocellulosic waste, ignored microbial community characterization analysis. This study aims to investigate whether the anaerobic microbial community can effectively degrade cellulose and hemicellulose present in lignocellulosic biomass. Addressing this question is essential for enhancing the fermentation performance of lignocellulosic biomass by improving the raw material utilization.

Table 9.1 Basic properties of substrates

Substrates	Total soild (TS, %)	Volatile solid (VS, %)	NH_3-N (mg/L)	Lignin (% TS)	Cellulose (% TS)	Hemicellulose (% TS)	C (%)	H (%)	N (%)	O (%)
GS	7.65	8.90	4.95	2.67	0.62	29.01	38.86	5.48	8.74	42.76
MC	95.90	95.90	–	0	100	0	43.39	6.41	0.01	49.09
XY	96.37	96.37	–	0	0	100	44.11	5.88	0	47.06
LI	96.98	55.20	–	100	0	0	–	–	–	–
CS	90.70	82.41	–	3.23	32.42	25.59	39.18	5.18	0.88	41.99

9.2 Materials and Methods

9.2.1 Substrates Characteristics

Microcrystalline cellulose (MC), xylan (XY), and alkaline lignin (LI) were used as substrates, and representing pure cellulose, hemicellulose, and lignin respectively. Corn straw (CS) was collected from a farm in Yanqing District, Beijing, and was air-dried, chopped, and then ground before use. Anaerobic granular sludge (GS) was obtained from a anaerobic plant in Weifang, Shandong. The anaerobic granular sludge was degassed in a water bath at 35 ± 1 °C before used as inoculum in anaerobic digestion. The basic properties of substrates are shown in Table 9.1.

9.2.2 Experimental Design

MultiTalent automatic methane potential test system (Nova Skantek (Beijing) Co.,Ltd) was used to evaluate the biomethane potential with 35 ± 1°C. When the methane production is less than 1% of the total production for 3 days, the anaerobic fermentation cycle automatically ends. 3 mol/L NaOH solution was used to absorb CO_2 in biochemical methane potential (BMP) testing instrument and mechanical stirring was performed at 100 rpm for 5 min every hour during fermentation cycle.

Microcrystalline cellulose, xylan, alkaline lignin, and corn straw were mixed with anaerobic granular sludge for anaerobic digestion (organic load rate 10 g VS/L). The ratio of inoculum to substrate was set to 2:1 (VS). The carbon-to-nitrogen ratio (25:1) and the initial pH of 7.5 was adjusted before the initiation of AD. The blank control groups (CK) were set up, and the cumulative methane production of the experimental group was calculated by subtracting the gas production of the blank control group. All test groups were established in triplicate, and the results were averaged for analysis.

9.2.3 Analytical Methods

9.2.3.1 Physicochemical Parameters Analysis.

The TS and VS were determined according to standard methods (Li et al. 2018b). Ammonia nitrogen was measured using water quality analyzer (HI83206, HANNA, Italy). Total carbon, nitrogen, hydrogen, and oxygen elements were determined through elemental analyzer (varioELcube, GmbH, Germany) as described in the literature. Lignin, cellulose, and hemicellulose content were determined using a chemical analyzer (F2000, Hanon, China).

9.2.3.2 Biomethane Analysis

The BMP testing instrument was used to determine the methane generation rate and methane yield, and the parameter value of methane content of biogas component in BMP equipment is set according to previous test data. Based on the elemental composition in Table 9.1, the theoretical methane production (TMP) was calculated using the total biochemical reaction equations for anaerobic digestion (Eqs. 9.1 and 9.2). The biodegradability (BD) of the raw materials was calculated using Eqs. (9.3).

$$C_nH_aO_bN_c + \left(n - \frac{a}{4} - \frac{b}{2} + \frac{3c}{4}\right)H_2O \rightarrow \left(\frac{n}{2} + \frac{a}{8} - \frac{b}{4} - \frac{3c}{8}\right) CH_4 + \left(\frac{n}{2} - \frac{a}{8} + \frac{b}{4} + \frac{3c}{8}\right)CO_2 + cNH_3 \tag{9.1}$$

$$TMP = \frac{22.4 \times 1000 \times \frac{n}{2} + \frac{a}{8} - \frac{b}{4} - \frac{3c}{8}}{12n + a + 16b + 14c} \tag{9.2}$$

$$BD = \frac{EMY}{TMP} \times 100\% \tag{9.3}$$

9.2.3.3 Kinetic Model.

This study employed modified Gompertz model (Eq. 9.4), Logistic model (Eq. 9.5), Cone model (Eq. 9.6), and first-order kinetic model (Eq. 9.7) to simulate the methane production from microcrystalline cellulose, xylan, lignin, and corn straw (Konan et al. 2024). Where B is the cumulative methane production (mL/g VS), B_0 is the maximum methane production in the experiment (mL/gVS), R_{max} is the maximum methane production rate (mL/gVS·d^{-1}), λ is the methane lag phase (d), k is the rate constant (d − 1), n is the shape coefficient, and t is the reaction time (d).

$$B = B_0 * \exp\left\{-\exp\left[\frac{R_{\max} * e}{B_0}(\lambda - t) + 1\right]\right\} \tag{9.4}$$

$$B = \frac{B_0}{1 + \exp\left[\frac{4R_{\max}}{B_0}(\lambda - t) + 2\right]} \tag{9.5}$$

$$B = \frac{B_0}{1 + (kt)^{-n}} \tag{9.6}$$

$$B = B_0 * \left[1 - \exp(-kt)\right] \tag{9.7}$$

9.2.3.4 Microbial Community Analysis

Samples were collected on the last fermentation day from all the anaerobic digestion reactors and stored at -80°C. 16S rRNA technology was used to interpret and discover microbial community characteristics during anaerobic fermentation of lignocellulosic biomass. In this study, the library was constructed by using the Ion Plus Fragment Library Kit 48 rxns (16S V3—V4) and quantified by Qubit and Q-PCR. The quantified library sequenced by using Thermofisher Ion S5TMXL. All samples were tested in triplicate times.

9.3 Results and Discussion

9.3.1 Methane Production

Agricultural residues such as lignocellulosic organic waste primarily contain lignin, cellulose, and hemicellulose. This study focused on biomethane production performance of these four fermentation substrates (lignin, cellulose, hemicellulose, and corn straw). By examining the gas production performance of different fermentation substrates, the trends in methane production from cellulose anaerobic fermentation were illustrated in Fig. 9.1. Methane was not produced in lignin system, mainly due to the stubborn physical structure of lignin, which is difficult for anaerobic bacteria to hydrolyze and utilize without chemical pretreatment, thermal treatment, or biological pretreatment.

XY system and CS system rapidly produced gas on the first day of fermentation. On the 10th day, the cumulative gas production of XY was below 1% total gas production. The cumulative methane production of XY system was up to 370.83 ± 14.60 mL/g VS. The cumulative methane production from CS was 172.06 ± 8.56 mL/g VS on 12th day. According to the test results, the fermentation cycle of XY was 10 days, while the straw was 12 days, and the cellulose system was 25 days. The MC

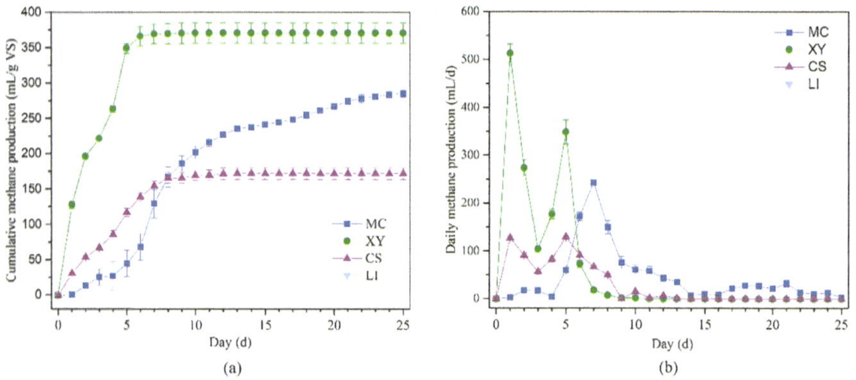

Fig. 9.1 Cumulative biomethane yield (**a**), and Daily methane production (**b**) of cellulose (MC), hemicellulose (XY), corn straw (CS) and lignin (LI)

system had slow gas production in the first three days, followed by a rapid increase after the 4th day. At the 25th day, the cumulative methane production stabilized (285.23 ± 5.17 mL/ VS), it was similar to the results of previous studies (Li et al. 2018a).

Compared to XY and CS, the MC anaerobic fermentation system had a lag phase of 3 to 4 days as shown in Fig. 9.1. On the 5th day, methane production rapidly increased and continued to rise. XY and CS system produced methane rapidly on the first day of anaerobic fermentation, indicating that hemicellulose can be quickly decomposed and digested by acid-producing bacteria in the anaerobic system. Corn straw contains 25.59% hemicellulose, which is why it can rapidly produce methane in the early stages of fermentation. However, its methane production rate was lower than that of the xylan fermentation system, because the particle size of corn straw (20 mesh) was much larger than that of xylan (100 mesh). Smaller particle sizes were more easily decomposed and digested by fermentation bacteria, and the decomposition of hemicellulose in corn straw was also affected by the complex structure of lignin, cellulose, and hemicellulose.

9.3.2 *Biodegradability*

Based on the elemental composition in Table 9.1, the theoretical methane production (TMP) was calculated using the total biochemical reaction equations for anaerobic digestion (Eqs. 9.1 and 9.2). The biodegradability (BD) of the raw materials was calculated using Eq. 9.3.

The theoretical unit VS methane production of MC was 416.84 mL/g VS, while the actual experimental unit VS methane production (EMY) was 285.23 mL/ VS. The theoretical unit VS methane production for xylan was 428.65 ml/g VS, while the actual unit VS methane production was 370.83 mL/g VS. Therefore, as shown in

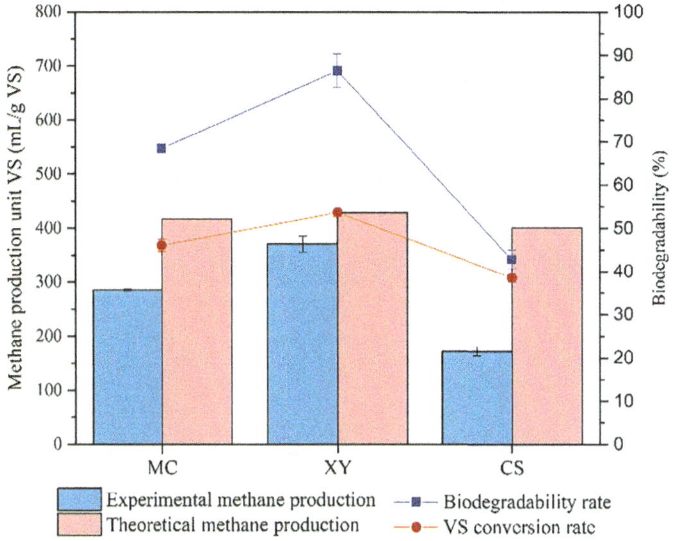

Fig. 9.2 Biodegradability of different compositions in lignocellulosic biomass

Fig. 9.2, the biodegradability of XY (86.51%) was higher than that of MC (68.43%). Corn straw, which was not pretreated before the methane generation potential study, had a lower biodegradability rate (only 42.91%). The methane production of MC was higher than that from corn straw, but still lower than xylan on VS conversion rate.

9.3.3 Kinetic Analysis

The first-order kinetic model is typically used to simulate the exponential phase of gas production for substrate hydrolysis rate constant. According to Table 9.2, the hydrolysis rate constant for the experimental microcrystalline cellulose was 0.0594 (typical value 0.04–0.13), while the highest hydrolysis rate was 0.3824 in XY anaerobic digestion system. The hydrolysis rate of corn straw, which contains 25.59% hemicellulose, was 0.2267. Based on R^2 and absolute percentage deviation values, the first-order kinetic model was not suitable for simulating the whole anaerobic fermentation process. The modified Gompertz model outperformed the first-order kinetic model, with an absolute percentage deviation of 4.84% for MC methane production. The modified Gompertz model is a typical "S"-shaped curve equation that can estimate methane production potential and gas production lag phase. The Logistic model is a classic growth function used to fit in gas production curves. Under the conditions of both models, the methane lag phase for the MC anaerobic system was the longest, at 3.16 days and 3.51 days, indicating that the hydrolysis

Table 9.2 Fitting parameters for cumulative methane production from different models

Fitting model		Modified gompertz	Logistic	Cone	First-order kinetic model
R^2	MC	0.9897	0.9844	0.9918	0.9394
	XY	0.9749	0.9971	0.9761	0.9782
	CS	0.9915	0.993	0.9787	0.9711
λ (d)	MC	3.1587	3.5107	–	–
	XY	0.000215	0.000277	–	–
	CS	0.3360	0.5187	–	–
k (d^{-1})	MC	–	–	0.1310	0.0594
	XY	–	–	0.5326	0.3824
	CS	–	–	0.2921	0.2267
Predicted methane yield (mL/g VS)	MC	271.43	265.02	274.68	301.70
	XY	372.09	370.63	379.35	375.11
	CS	173.57	172.37	176.60	177.28

and acidification phase was the main limiting period for cellulose anaerobic fermentation. Moreover, Cone model can effectively simulate the gas production initiation and exponential phases. Cone model had the best simulation effect for MC anaerobic fermentation, with the highest correlation coefficient (R2 = 99.18%) as shown in Table 9.2.

9.3.4 Microbial Community Characteristics

As shown in Fig. 9.3, *T78*, *vadinCA02*, *Clostridium* and *Bacteroides* existed in the MC, XY and CS anaerobic digestion systems at the same time, indicating that these were involved in the hydrolysis and acidification process of lignocellulosic biomass components. *T78* exhibited microbial indicators attribute, and played important roles in hydrolytic-acidification (Cortez-Cervantes et al. 2024). *Clostridium*, belonging to Firmicutes, is a type of bacteria that can decompose cellulose under anaerobic conditions. After anaerobic fermentation, Compared to CK (0.95%), the relative abundance of *Clostridium* increased to 5.65% (MC), 3.65% (XY) and 2.84% (CS) respectively.

Notably, the relative abundance of *Methanosaeta* was highest, indicating that the reaction system was primarily acetate-type fermentation. Additionally, *Methanobacterium* also had higher relative abundance. After the anaerobic fermentation of MC, XY, and CS, the relative abundance of *Methanospirillum* increased, which was the facultative methanogen (Jung et al. 2022). Although the relative abundance of *Methanosaeta* and *Methanobacterium* were at relatively high levels, biomethane did not produced in the LI system.

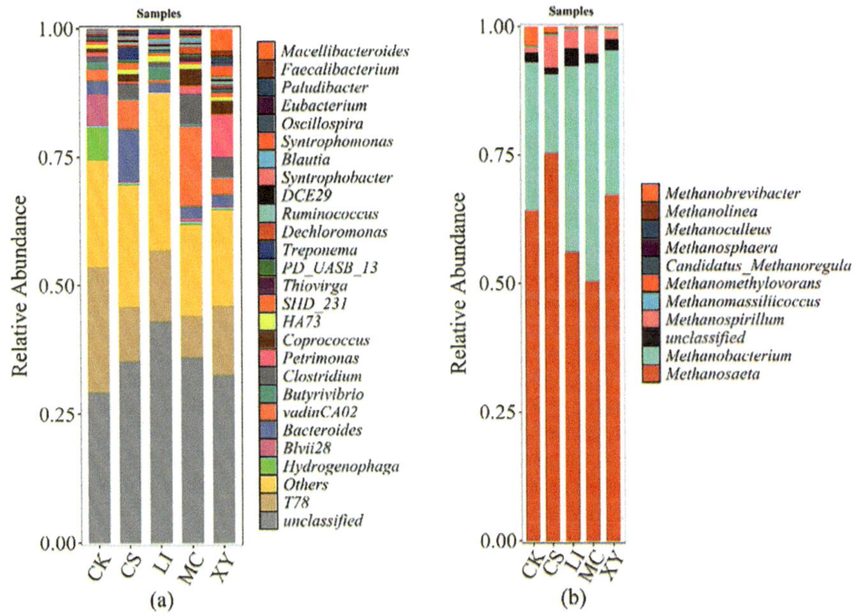

Fig. 9.3 Taxonomic distribution of bacterial (**a**) and archaeal (**b**) sequences at the genus level

9.4 Conclusions

The production potential of biomethane from different components of lignocellulosic biomass was studied. The cumulative methane production of MC, XY, and CS system was 285.23 ± 5.17 mL/ VS, 370.83 ± 14.60 mL/g VS, 172.06 ± 8.56 mL/g VS respectively. Methane was not produced in lignin system. Cellulose and hemicellulose can be converted by microorganisms into biomethane, and the biodegradability rate VS conversion rate were higher than that from corn straw. In addition, the fermentation cycle of MC (25 d) anaerobic system was longer than XY (10 d) and CS (12d). The lag time of MC system was longer than XY, and CS system. The results of dynamic model also revealed this point. Lignocellulosic biomass anaerobic digestion had special microbial community characteristics. *T78*, *vadinCA02*, *Clostridium* and *Bacteroides* played important roles in hydrolytic-acidification and existed in the MC, XY and CS anaerobic digestion systems at the same time. The relative abundance of *Clostridium*, which can decompose cellulose under anaerobic condition, increased to 5.65% (MC), 3.65% (XY) and 2.84% (CS) respectively after anaerobic fermentation. *Methanosaeta* and *Methanobacterium* were the two main anaerobic archaea in MC, XY and CS anaerobic digestion systems.

References

Cortez-Cervantes J, Moreno-Andrade I, Escalante AE, de los Cobos-vasconcelos D, Carrillo-Reyes J (2024) Identifying reliable microbial indicators in anaerobic digestion of organic solid waste: Insights from a meta-analysis. J Environ Chem Eng 12(5)

Du J, Qian Y, Xi Y, Lü X (2019) Hydrothermal and alkaline thermal pretreatment at mild temperature in solid state for physicochemical properties and biogas production from anaerobic digestion of rice straw. Renew Energy 139:261–267

Jung S, Kim M, Lee J, Shin J, Shin SG, Lee J (2022) Effect of magnetite supplementation on mesophilic anaerobic digestion of phenol and benzoate: Methane production rate and microbial communities. Bioresource Technology 350

Isikgor FH, Becer CR (2015) Lignocellulosic biomass: a sustainable platform for the production of bio-based chemicals and polymers. Polym Chem 6:4497–4559

Kucharska K, Hołowacz I, Konopacka-Łyskawa D, Rybarczyk P, Kamiński M (2018) Key issues in modeling and optimization of lignocellulosic biomass fermentative conversion to gaseous biofuels. Renew Energy 129:384–408

Konan D, Rodrigue D, Koffi E, Elkoun S, Ndao A, Adjallé K (2024) Combination of technologies for biomass pretreatment: a focus on extrusion. Waste and Biomass Valorization. 15(8):4519–4540

Li WW, Khalid H, Zhu Z et al (2018a) Methane production through anaerobic digestion: Participation and digestion characteristics of cellulose, hemicellulose and lignin. Appl Energy 226:1219–1228

Li Z, Chen Z, Ye H, Wang Y, Luo W, Chang J-S, Li Q, He N (2018b) Anaerobic co-digestion of sewage sludge and food waste for hydrogen and VFA production with microbial community analysis. Waste Manage 78:789–799

Wang L, Cao Z, Zou J, Liu Z, Li Y, Wang Z (2020) Urea-pretreated corn stover: Physicochemical characteristics, delignification kinetics, and methane production. Biores Technol 306:123097

Wei Y, Li Z, Ran W, Yuan H, Li X (2021) Performance and microbial community dynamics in anaerobic co-digestion of chicken manure and corn stover with different modification methods and trace element supplementation strategy. Bioresource Technolog 325:124713

Chapter 10
Experimental Study on Water–Rock Coupling of Red Sandstone

Juntao Zhu, Xiangmei Chen, Yongqiang Ren, Qi Liu, Zhihui Mi, Yingchun He, and Wenlong Cao

Abstract To elucidate the impact of wet-dry cycles on red sandstone in the Ordos region, experiments were conducted using a servo-hydraulic machine. Parameters such as uniaxial compressive strength, elastic modulus, strain corresponding to peak stress, and Poisson's ratio were evaluated across various cycles. The study reveals a significant decline in uniaxial compressive strength, elastic modulus, and peak strain of red sandstone under wet-dry cycling. However, the effect on Poisson's ratio remains minor. The findings provide valuable insights for engineering design and slope failure mechanism research.

Keywords Rockmechanics · Dry and wetcycle · Peakstrain · Poissons ratio · Uniaxial compressive strength · Elastic modulus

10.1 Introduction

The mechanical properties of sandstone are not only related to the mineral composition of the rock and the development of the structural surface, but also related to the environment of the rock and the external conditions such as the load. The mechanical properties reflected by the rock are the result of the coupling of internal and external factors. In recent years, with more and more infrastructure construction projects in China, we are facing more and more problems, among which the main problems are

J. Zhu · X. Chen (✉) · Y. Ren · Q. Liu · Z. Mi · W. Cao
Department of Civil Engineering, Ordos Institute of Technology, Ordos, China
e-mail: geochem_chen19@oit.edu.cn

X. Chen
College of Chemical Engineering, Inner Mongolia University of Technology, Hohhot, China

J. Zhu
Inner Mongolia University of Science and Technology, Baotou, China

Y. He
Inner Mongolia Institute of Technology, China University of Geosciences, Ordos, China

Q. Wang and A. Hosseini-Bandegharaei (eds.), *Environmental Functional Materials in Air Pollution*, Environmental Science and Engineering,
https://doi.org/10.1007/978-3-031-84571-0_10

that the erosion of groundwater leads to the expansion and softening of sandstone, and the water–rock-dynamic coupling of sandstone has become an urgent problem to be studied and solved in the development of infrastructure construction projects.

currently, Chinese scholars have done relatively little research on this aspect (Chen et al. 2024; Ren and Chen 2024), The analysis of the influencing factors of the existing studies is not too detailed, For example, the influence of dry and wet cycle on the impact dynamic properties of clay slate in the study of water–rock coupling phenomenon of slite; Wang Laigui et al. (Chen et al. 2021) in the different solution environment of sandstone dry and wet cycle damage characteristics test research wrote in the softening of sandstone mechanical properties under dry and wet cycle conditions; Meisheng et al. (Chen et al. 2024). Test and model study on the damage effect of sandstone under different solutions and dry and wet cycle studied the damage effect of acidic environment and dry and wet cycle on sandstone; Yanlin et al. (Laigui et al. 2020), in the softening and disintegration characteristics of muddy red sandstone under the action of dry and wet cycle, studied the widely distributed red sandstone in South China for the dry and wet cycle test and detected the pore characteristics of the sample during the dry and wet cycle, The damage mechanism of the red sandstone during the dry and wet cycle is analyzed. At present, most of the existing studies consider single factors, and the few studies of water–rock-dynamic coupling are not aimed at red sandstone. Therefore, the current research results (Meisheng et al. 2024; Liu et al. 2020; Sun et al. 2023) are difficult to meet the engineering practice requirements required for social development.

Taking this as the starting point, the direct shear experiment of red sandstone under different dry and wet cycles based on the three axis test, and then compares the test results under different dry and wet cycles, analyzes the strength variation law of sandstone, and explores the failure mechanism of sandstone slope and other projects. Some relevant conclusions are drawn to provide guidance for engineering design and research of slope failure mechanism.

10.2 Study Area and Methods

10.2.1 Study Area

The sample used in the experiment is the red sandstone in ordos area, the sampling site as shown in Fig. 10.1., the mineral composition of quartz, feldar, calcite, illite, montmorillonite, chlorite, hematite, etc., including montmorillonite is the main factors affecting the effect of water–rock coupling, it in the red sandstone is 1%. Red sandstone sample size 105 cm cylindrical sample.

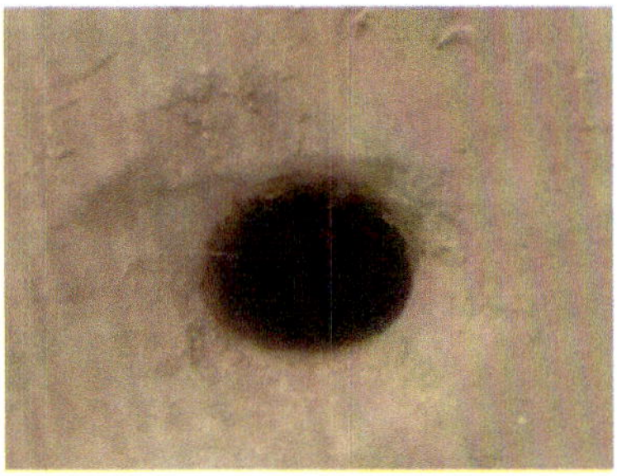

Fig. 10.1 Sampling location

10.2.2 Methods

Based on the "Code for Rock Tests in Road Engineering" (JTGE41-2005), laboratory tests were conducted to determine the basic physical properties of red sandstone. The uniaxial compression test of red sandstone was conducted using the new testing machine (TSW-1000) from Changchun, as shown in Fig. 10.2. This machine, manufactured by Changchun New Testing Machine Co., Ltd., can apply a maximum axial force of 2000 kN and a maximum confining pressure of 100 MPa. It is capable of performing standard unconfined compression tests and triaxial compression tests.

Fig. 10.2 Rock triaxial experiment system

Table 10.1 Main physical water properties parameters of rock samples

Formation lithology	Natural density ρ (g/cm^3)	Dry density ρ d (g/cm^3)	Saturation density ρ w (g/cm^3)	Saturation water absorption rate W sa (%)	Natural water absorption rate W a (%)	natural moisture content W (%)	Surfilled water system K s
Red sandstone	2.59	2.58	2.64	4.42	2.55	0.51	0.58

10.3 Results and Discussion

10.3.1 The Fundamental Physics of Red Sandstone

Samples of 6 to 12 representative red sandstone specimens undergo testing for bulk density, moisture content, and water absorption. Results are shown in Table 10.1. The formulae (1) and (2) allow for the determination and computation of the bulk density of red sandstone.

$$\rho_d = \frac{m_s}{\frac{m_1 - m_2}{\rho_w} - \frac{m_1 - m_2}{\rho_p}} \tag{10.1}$$

$$\rho = \frac{m_s}{\frac{m_1 - m_2}{\rho} - \frac{m_1 - m_2}{\rho_p}} \tag{10.2}$$

where,

ρ—wet density of rock mass (g/cm3);
m—wet Specimen (g);
$m1$—wax Seal Wet Specimen (g);
$m2$—wet Specimen in Water (g);
ρ_p—Paraffin Density (g/cm^3).

10.3.2 The Mechanical Properties of Red Sandstone

To minimize the dispersion in test results, average the values from three samples after each test. Use displacement control mode with a loading rate of 0.02 mm/s until sample failure and then release pressure. When studying the mechanical properties of red sandstone, this paper calculates the arithmetic mean of each group's test results to reduce data dispersion due to rock anisotropy. Table 10.2 presents the test results, and Fig. 10.3 shows the tested red sandstone.

Table 10.2 List of uniaxial test results of red sandstone

Test parameters	Rock-like state	
	Natural air dry	saturate
Single axis compressive strength (MPa)	45.46 –60.78 (55.38)	35.55–38.5 (35.98)
Elassitive modulus E (GPa)	13.13–20.45 (14.77)	6.94–9.39 (8.57)
Poisson ratio μ	0.15–0.33 (0.21)	0.12–0.26 (0.20)
Range of wave speed values (m/s)	4415–4674	4620 –4693

Fig. 10.3 Soil samples after the test

10.3.3 Analysis of Water Absorption Test Law of Red Sandstone Under the Coupling Action of Water Rocks

Based on the results of water absorption tests on rock samples for different cycle counts, using the natural moisture content as the initial baseline, the absorption rate after 25 cycles is recorded, as shown in Tables 10.3 and 10.4.

We assume that the water content of the sample is continuous within five cycles, so we can analyze the water content of Tables 10.4 and 10.5, and obtain the water absorption rate of soft rocks and the number of dry and wet cycles as shown in Fig. 10.4.

It can be seen from the chart that the actual water absorption rate of soft rock ω N, ω N = 0.6515 ln (N) + 1.9979, $R^2 = 0.9567$. The general trend of the curve

Table 10.3 The absorbent water absorption test results of red sandstone

Cycle index	Natural state mass (g)	Air-dry state mass (g)	Saturation state mass (g)	Actual water absorption of (%)
1	475.6	465.99	–	2.02
5	–	467.76	482.32	3.11
10	–	467.84	482.82	3.21
15	–	467.05	484.2	3.94
25	–	467.63	486.9	4.12

Table 10.4 Average increment of water absorption in red sandstone

Cycle stage (i)	cycle index (N)	Actual water absorption rate in N cycles (%)	Stage average water absorption (g)$\overline{M_{abs}}$	Stage average water absorption increment (g)$\overline{\Delta M_{abs}}$
I (1)	1	2.02	8.77	8.77
II (4)	5	3.11	13.33	1.52
III (5)	10	3.21	15.08	0.35
IV (5)	15	3.94	17.63	0.37
V (10)	25	4.12	17.81	0.02

Table 10.5 Table captions should be placed above the tables Uniaxial compression results of red sandstone after hydrorock coupling

Test team	Recurrence stage	Uniaxial compressive strength of σc (MPa)		Modulusof elasticity E (GPa)		Peak strain εc(× 10^{-3})		Poisson ratio μ	
		Trial value	Average value	Trial value	Average value	Trial value	Average value	Trial value	Average value
I	0 Circles	33.5	32.41	9.392	9.20	4.564	4.51	0.12	0.18
		35.79		9.415		4.443		0.26	
		27.96		8.803		4.524		0.16	
II	5 cycles	45.42	29.95	10.95	9.30	4.84	3.80	0.15	0.218
		32.48		8.998		4.14		0.27	
		17.96		7.959		2.42		0.22	
III	10 cycles	28.784	27.90	7.308	7.45	5.307	5.07	0.13	0.15
		32.745		8.748		4.559		0.21	
		22.174		6.28		5.338		0.12	
IV	15 Circles	30.883	27.05	7.441	7.51	5.019	4.37	0.23	0.17
		33.509		8.851		4.866		0.10	
		16.764		6.246		3.231		0.17	
V	25 Circles	21.548	17.30	5.581	4.77	4.739	4.38	0.11	0.16

shows that the water absorption of soft rock increases with the number of dry and wet cycles, especially in the stage of early drying to re-water absorption, and tends to the late stage.

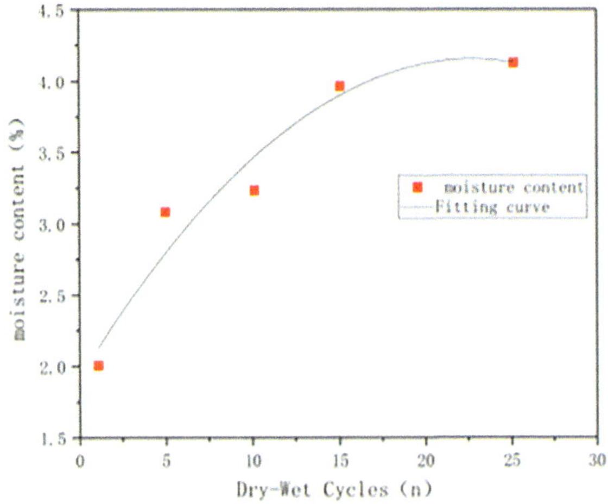

Fig. 10.4 Relationship between dry and wet cycle times and water absorption rate of rock

10.3.4 Results of Uniaxial Compressive Test of Red Sandstone Under the Coupled Action of Water Rock

The results of the experimental tests are shown in Table 10.5. According to Fig. 10.5, the relationship between the dry and wet cycle times and the uniaxial compressive strength, the more the dry and wet cycles, the lower the uniaxial compressive strength of the rock is reduced, until the rock mass is destroyed, and the compressive strength of the rock mass is suddenly reduced.

By Fig. 10.6 dry wet cycle frequency and the elastic modulus figure, the elastic modulus of the rock before five dry wet cycle change is not obvious, five to ten dry

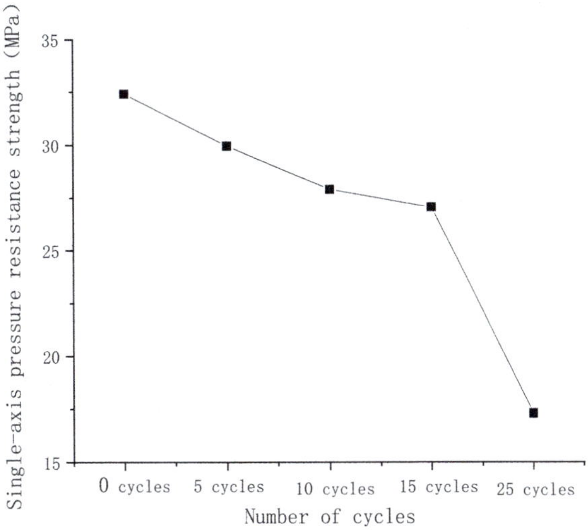

Fig. 10.5 The relationship between the number of dry and wet cycles and the uniaxial compressive strength

wet cycle in the tenth and fifteen dry and wet cycle change is not obvious, finally the rock is destroyed, elastic modulus drop, thus the dry wet cycle will reduce the elastic modulus of red sandstone.

As can be seen from the dry and wet cycle times and peak strain curve (Fig. 10.7), with the increase of dry and wet cycles, the sandstone cracks increase and the peak force gradually decreases; the fifth to the tenth cycle may produce new minerals or cements to increase the peak force, the peak force of no new material formation gradually decreases, and the damaged peak strain of the 15th cycle sandstone remains basically unchanged.

According to the relationship between the number of dry and wet cycles and Poissons ratio in Fig. 10.8, it is the most sensitive to the uniformity of rock, while

Fig. 10.6 Number of dry and wet cycles and elastic modulus

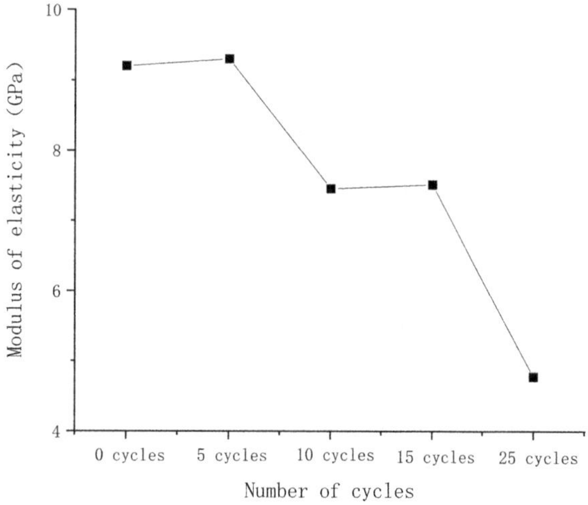

Fig. 10.7 Number of dry and wet cycles versus peak strain

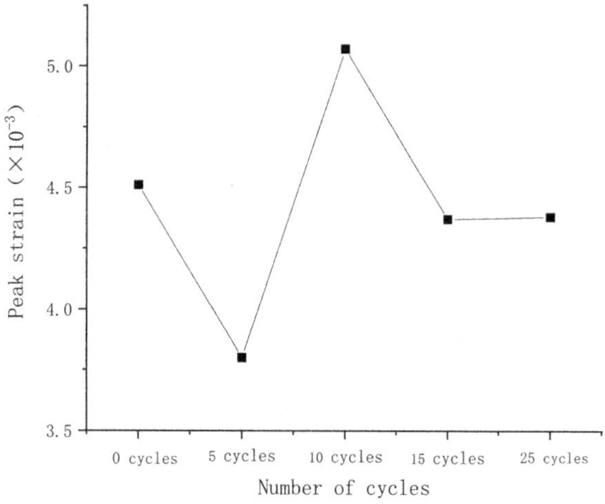

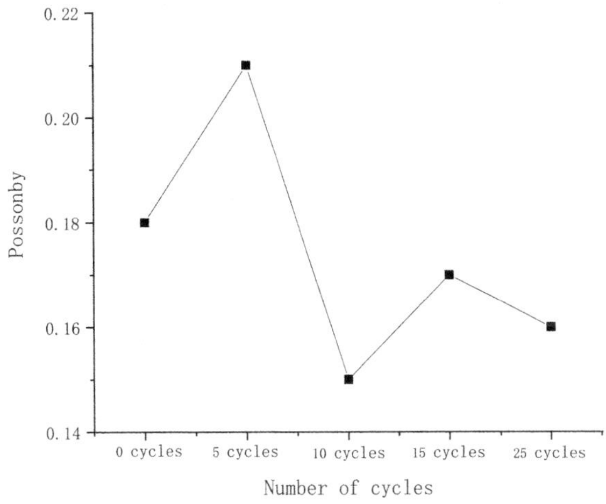

Fig. 10.8 Dry and wet cycle number and Poissons ratio

red sandstone is a kind of uneven rock. Under the action of dry and wet cycle, the cracks inside the rock will gradually develop, which will eventually lead to the destruction of the rock. Therefore, the dry and wet cycle makes the red sandstone inhomogeneity more obvious, and there is no obvious change between the dry and wet cycle times and Poissons ratio.

According to the correlation analysis between σ c and E and dry and wet cycle number n, their change trend is strongly correlated with the cycle number n, and they are all subject to a negative exponential distribution.

10.4 Conclusions

This paper through the experimental analysis of sandstone water absorption, based on Changchun new test machine (TSW-1000), the sandstone under different water absorption impact compression experiment, then compare the experimental results under different water content, the analysis of the sandstone strength change rule, analyzes the water on sandstone and influence effect, the red sand in the engineering design and slope engineering has certain reference value, can draw the following conclusions:

1. The water absorption rate of soft rock increases with the increase of dry and wet cycles, especially in the initial drying to re-water absorption phase, with the increase of dry and wet cycles, the faster the water absorption rate of soft rock, the closer to the saturation state
2. Under the action of dry and wet cycle, the uniaxial compressive strength of the rock gradually decreases until the rock mass is destroyed, and the compressive strength of the rock mass drops suddenly.
3. Dry and wet cycles will reduce the elastic modulus of the red sandstone.

4. As the number of dry and wet cycles increases, the crack in the sandstone becomes larger, the force reaching the peak gradually decreases, and the peak strain is basically kept unchanged when the sandstone is damaged.
5. The dry and wet cycle makes the homogeneity of red sandstone more obvious, and there is no obvious change between the dry and wet cycle times and Poissons ratio.

References

Chen X, Ren Y, Tang B et al (2021) Stability analysis by finite element method and treatment measures of slide slope in open-pit coal mine. In: Proceedings of SPIE—The International Society for Optical Engineering12050

Chen X, Ren Y, Tang B et al (2024) Influence of mechanical and microscopic properties of red sandstone modified by different solid waste materials. Adv Civ Eng

Chen X, Ren Y, Tang B et al (2024) Experimental study of macro- and micro-scopic damage inred sandstone under dry and wet cycling. Adv Civ Eng

Laigui W, Shengpeng D, Wenbo A, Qiang C (2020) Experimental study on sandstone dry and wet cycle damage characteristics under different solution environment. Exper Mech 35(03):521–531

Liu Q, Sun Y, Li J (2020) Experimental study on seepage characteristics of Jurassic weakly cemented sandstone under water-rock interaction. Geofluids 2020(1468–8115):1–12

Meisheng F, Qianhua Y, Xinyu Z (2024) Test and model of sandstone under different solutions and dry and wet cycles 52(01):20–26+33

Ren Y, Chen X, Shang Y (2024) Study on the influence of rainfall infiltration on the stability of dump slope. Desal Water Treatment 319

Sun X et al (2023) Study on microscopic mechanism and failure mode of water-rock coupling in fine sandstone under uniaxial compression. Rock Mech Rock Eng 56(0723–2632):6823–6842

The manufacturer's authorised representative in the EU is Springer Nature Customer Service Centre GmbH, Europaplatz 3, 69115 Heidelberg, Germany. If you have any concerns regarding our products, please contact ProductSafety@springernature.com

Printed and bound by CPI Group (UK) Ltd, Croydon, CR0 4YY
30/09/2026
02225604-0002